AF327624

Takashi Sakurai

Editor

The Hinode Mission

Chapters previously published in *Solar Physics* Volume 243, Issue 1, 2007 or Volume 249, Issue 2, 2008

 Springer

Takashi Sakurai
National Astronomical Observatory of Japan
Tokyo, Japan

Cover illustration: Courtesy of ISAS/JAXA

© Japan Aerospace Exploration Agency, 2008

All rights reserved.

Library of Congress Control Number: 2008938950

ISBN-978-0-387-88738-8   e-ISBN-978-0-387-88739-5

Printed on acid-free paper.

© 2008 Springer Science+Business Media, BV

No part of this work may be reproduced, stored in a retrieval system, or transmitted in any form or by any means, electronic, mechanical, photocopying, microfilming, recording or otherwise, without the written permission from the Publisher, with the exception of any material supplied specifically for the purpose of being entered and executed on a computer system, for the exclusive use by the purchaser of the work.

1

springer.com

Contents

In Memoriam: Takeo Kosugi

Originally published in the journal Solar Physics, Volume 243, No 1.
DOI: 10.1007/s11207-007-9012-8 © Springer Science+Business Media B.V. 2007

Takeo Kosugi (1949–2006)

Takeo Kosugi, project manager of the *Hinode* mission, died suddenly on 26 November 2006, from a cerebral infarction. He was 57 years old.

Takeo was born in 1949 in Toyohashi city near Nagoya. After graduating from the University of Tokyo in 1972 he proceeded to the graduate course in astronomy there, and in

1976, while he was in the Ph.D. course, he was appointed as research associate at the Nobeyama Solar Radio Observatory of the Tokyo Astronomical Observatory, the University of Tokyo.

His initial research activities were based at Nobeyama where the main facilities were a two-dimensional interferometer (but with separate $E-W$ and $N-S$ systems) at 160 MHz and a 17 GHz interferometer. With his senior colleagues, Hiroshi Nakajima and the late Keizo Kai, he participated in building an opto-acoustic spectrometer and in upgrading the 17 GHz interferometer. He completed his Ph.D. thesis in 1984 on the directivity of radio emission from solar flares using the 17 GHz data.

When the Astro-A (*Hinotori*) satellite was launched in 1981, he applied his radio-astronomy skills, namely image synthesis based on CLEAN, to hard X-ray imaging. This was the beginning of his involvement in space solar astronomy. In the 1980s a plan for building a dedicated solar array at microwave frequencies emerged in the Japanese solar radio-astronomy community, and he was a key member in its initial phase. Eventually, the project materialized as the Nobeyama Radioheliograph, with first light in 1992, but his main field had shifted to the space program. In 1988 he was promoted to associate professor, and in 1992 he moved to the National Astronomical Observatory of Japan as professor.

When the Solar-A satellite program began, he joined the team for a hard X-ray telescope (HXT), and later he was appointed principal investigator for HXT. Solar-A was renamed *Yohkoh* after its launch in 1991, and the combination of HXT with a soft X-ray telescope (SXT) and an X-ray spectrometer (BCS) led to a very successful mission. In particular, the discovery of a new type of coronal hard X-ray source with S. Masuda, then a student of Takeo, opened an important avenue in studying particle acceleration in solar flares. He also assisted the *Yohkoh* project manager Yoshiaki Ogawara, and he himself served as project manager since 1998, in operating the *Yohkoh* satellite until its reentry into the atmosphere in 2005.

When the next solar mission, Solar-B, was approved, he moved to ISAS (the Institute of Space and Astronautical Science, which later became ISAS/JAXA), to take the role of project manager in 1998. The satellite was successfully launched on 23 September 2006, and was given the name *Hinode* (Sunrise) by him. His face – full of joy and confidence at the press conference just after launch – was remembered by all who participated in the Solar-B project. The press release with the initial scientific data from the optical telescope (SOT) onboard *Hinode* took place on 27 November, but he had passed away suddenly on the previous day. Accumulation of fatigue from his stressful life might have been a contributing factor.

His talent in organizing a large number of people and in leading a big project was extraordinary, and therefore he had been actively involved in many international as well as Japanese organizations, including COSPAR, the International Living with a Star program, CAWSES, and IHY. He was science director of ISAS/JAXA from October 2005 and had been a member of numerous committees. He had wide media exposure and served as spokesperson of solar and space physics. He had been recording an educational lecture for television on 24 November and collapsed late at night after returning home.

He is survived by his wife Kikuko, by two sons and a daughter, and by *Hinode*.

5 March 2007
Takashi Sakurai
Taro Sakao
Hugh S. Hudson

Preface

The *Solar-B* satellite was launched in the morning of 23 September 2006 (06:36 Japan time) by the Institute of Space and Astronautical Science, Japan Aerospace Exploration Agency (ISAS/JAXA), and was renamed to *Hinode* ('sunrise' in Japanese). *Hinode* carries three instruments; the X-ray telescope (XRT), the EUV imaging spectrometer (EIS), and the solar optical telescope (SOT). These instruments were developed by ISAS/JAXA in cooperation with the National Astronomical Observatory of Japan as domestic partner, and NASA and the Science and Technology Facilities Council (UK) as international partners. ESA and Norwegian Space Center have been providing a downlink station. All the data taken with *Hinode* are open to everyone since May 2007.

This volume combines the first set of instrumental papers of the *Hinode* mission (the mission overview, EIS, XRT, and the database system) published in volume 243, Number 1 (June 2007), and the second set of papers (four papers on SOT and one paper on XRT) published in Volume 249, Number 2 (June 2008). Another SOT paper cited as Tarbell *et al.* (2008) in these papers will appear later in Solar Physics.

Editor: Takashi Sakurai, National Astronomical Observatory of Japan, Mitaka, Tokyo, Japan

DOI: 10.1007/978-0-387-88739-5_2 © Springer Science+Business Media B.V. 2008

The *Hinode* (Solar-B) Mission: An Overview

T. Kosugi · K. Matsuzaki · T. Sakao · T. Shimizu · Y. Sone · S. Tachikawa ·
T. Hashimoto · K. Minesugi · A. Ohnishi · T. Yamada · S. Tsuneta · H. Hara ·
K. Ichimoto · Y. Suematsu · M. Shimojo · T. Watanabe · S. Shimada · J.M. Davis ·
L.D. Hill · J.K. Owens · A.M. Title · J.L. Culhane · L.K. Harra · G.A. Doschek ·
L. Golub

Originally published in the journal Solar Physics, Volume 243, No 1.
DOI: 10.1007/s11207-007-9014-6 © Springer Science+Business Media B.V. 2007

Abstract The *Hinode* satellite (formerly *Solar-B*) of the Japan Aerospace Exploration Agency's Institute of Space and Astronautical Science (ISAS/JAXA) was successfully launched in September 2006. As the successor to the *Yohkoh* mission, it aims to understand how magnetic energy gets transferred from the photosphere to the upper atmosphere and results in explosive energy releases. *Hinode* is an observatory style mission, with all the instruments being designed and built to work together to address the science aims. There

T. Kosugi deceased 26 November 2006.

T. Kosugi · K. Matsuzaki · T. Sakao · T. Shimizu (✉) · Y. Sone · S. Tachikawa · T. Hashimoto ·
K. Minesugi · A. Ohnishi · T. Yamada
Institute of Space and Astronautical Science, Japan Aerospace Exploration Agency, Sagamihara,
Kanagawa 229-8510, Japan
e-mail: shimizu@solar.isas.jaxa.jp

S. Tsuneta · H. Hara · K. Ichimoto · Y. Suematsu · M. Shimojo · T. Watanabe
National Astronomical Observatory of Japan, Mitaka, Tokyo 181-8588, Japan

S. Shimada
Kamakura Works, Mitsubishi Electric Corp., Kamakura, Kanagawa 247-8520, Japan

J.M. Davis · L.D. Hill · J.K. Owens
Space Science Office, VP62, NASA Marshall Space Flight Center, Huntsville, AL 35812, USA

A.M. Title
Lockheed Martin Solar and Astrophysics Laboratory, B/252, 3251 Hanover Street, Palo Alto,
CA 94304, USA

J.L. Culhane · L.K. Harra
UCL Mullard Space Science Laboratory, Holmbury St. Mary, Dorking, Surrey RH5 6NT, UK

G.A. Doschek
E. O. Hulburt Center for Space Research, Code 7670, Naval Research Laboratory, Washington,
DC 20375-5352, USA

L. Golub
Smithsonian Astrophysical Observatory, Cambridge, MA 02138, USA

are three instruments onboard: the Solar Optical Telescope (SOT), the EUV Imaging Spectrometer (EIS), and the X-Ray Telescope (XRT). This paper provides an overview of the mission, detailing the satellite, the scientific payload, and operations. It will conclude with discussions on how the international science community can participate in the analysis of the mission data.

1. Introduction

The *Hinode* spacecraft (formerly Solar-B) of the Institute of Space and Astronautical Science, Japan Aerospace Exploration Agency (ISAS/JAXA), was launched on 22 September 2006, at 21:36 GMT, aboard the seventh in JAXA's series of M-V rockets. The principal scientific goals of the *Hinode* mission are the following:

(1) To understand the processes of magnetic field generation and transport including the magnetic modulation of the Sun's luminosity.
(2) To investigate the processes responsible for energy transfer from the photosphere to the corona and for the heating and structuring of the chromosphere and the corona.
(3) To determine the mechanisms responsible for eruptive phenomena, such as flares and coronal mass ejections, and understand these phenomena in the context of the space weather of the Sun – Earth System.

This mission is the follow-on to *Yohkoh,* an ISAS mission with significant NASA and United Kingdom participation that was launched in 1991 (Ogawara *et al.*, 1991) and continued taking observations for nearly a solar cycle. *Yohkoh* demonstrated that the high-temperature corona is highly structured and dynamic and that rapid heating and mass acceleration are common phenomena (Acton *et al.*, 1992). *Yohkoh* was launched shortly after the maximum of solar cycle 22, which was an ideal period for studying large solar flares. The subsequent observations provided considerable evidence to support magnetic reconnection as the driver for energy release in flares. Hard X-ray "above the loop top" sources were found in compact flares (*e.g.*, Masuda *et al.*, 1994) and also in long-duration flares (*e.g.*, Harra *et al.*, 1998). In soft X rays the flaring loops often took on the appearance of cusps, which is to be expected from the standard model where the reconnection occurs high in the corona (*e.g.*, Tsuneta, 1996; Canfield, Hudson, and McKenzie, 1999; Sterling *et al.*, 2000). The edges of the loops were also found to be hotter, as expected if the outer edges are the last to be heated from reconnection. As expected from the reconnection, plasma ejections from flaring sites have been found on many occasions (*e.g.*, Shibata *et al.*, 1995). On smaller scales, many jets were found in soft X rays; these are interpreted as reconnection occurring through the interaction of emerging flux and already existing magnetic field (Shimojo *et al.*, 1996). Many small-scale flares were observed in active region loops (*e.g.*, Shimizu, 1995; Shimizu *et al.*, 2002) and in bright points (Priest *et al.*, 1994). On a more global scale, dramatic coronal waves were observed (*e.g.*, Hudson *et al.*, 2003) and trans-equatorial loops were found to erupt (*e.g.*, Khan and Hudson, 2000) followed by coronal mass ejections or flares (Harra, Matthews, and van Driel-Gesztelyi, 2003).

Hinode is designed to address the fundamental question of how magnetic fields interact with the ionized atmosphere to produce solar variability. Measuring the properties of the Sun's magnetic field is the fundamental observational goal of *Hinode* and differentiates it from previous solar missions. The three instruments were selected to observe the response of the chromosphere and corona to changes in the photospheric magnetic field. To achieve this end *Hinode* makes quantitative measurements of all three components of vector magnetic fields. This allows calculation of the free energy of the magnetic field, which powers solar activity through the action of electric currents. The components of the magnetic field are difficult to resolve, especially from the ground where seeing effects degrade spatial resolution. The major scientific instrument on *Hinode*, the Solar Optical Telescope (SOT), makes these observations from space. The response of the solar atmosphere to magnetic field changes is measured by the EUV Imaging Spectrometer (EIS) and the X-Ray Telescope (XRT).

Based upon this scientific motivation, *Hinode* was planned and constructed as an international collaborative project including institutions in Japan, the United States, and the United Kingdom. ISAS/JAXA has responsibility for the design, development, test, and integration of the *Hinode* spacecraft with the National Astronomical Observatory of Japan as a domestic partner and the Mitsubishi Electric Corporation as a leading contractor. The participating institutes and their responsibilities are shown in Table 1. The *Hinode* spacecraft was called by its development name Solar-B and the name *Hinode* was given after successful launch according to the Japanese satellite tradition. *Hinode* is a Japanese word meaning sunrise.

In the present paper, we will give an overview of the *Hinode* mission from the viewpoints of the scientific instruments in Section 2, the spacecraft design in Section 3, and the flight operations in Section 4. The scientific objectives will be briefly discussed in Section 5.

2. Scientific Instruments

The scientific payload consists of three instruments: the SOT, the EIS, and the XRT. Each instrument is a result of the combined talents of all the members of the international team. Full technical details of each instrument are described in the separate papers in this special issue. This paper provides a brief summary of each instrument with their main characteristics summarized in Table 2. The instruments usually work together as an "observatory" studying the same target at which the spacecraft is pointed. Optionally, the EIS has the ability to offset its own pointing and the XRT, having a larger field of view than the others, has the ability to observe its own region of interest.

2.1. Solar Optical Telescope

The SOT is the largest solar optical telescope flown in space (Tsuneta *et al.*, 2007). The SOT consists of the Optical Telescope Assembly (OTA) (Suematsu *et al.*, 2007) and its Focal Plane Package (FPP) (Tarbell *et al.*, 2007). The OTA is a 50-cm clear aperture, aplanatic Gregorian, $f/9$ design telescope. The OTA is diffraction limited ($0.2'' - 0.3''$) between 3,880 and 6,700 Å. The primary mirror is fabricated out of ULE and supported by invar/titanium structures to retain thermal stability. Field stops and heat rejection mirrors are located at the focus of the primary mirror and at the Gregorian focus. The secondary field stop limits the field of view to $361'' \times 197''$. The OTA holds the collimating lens unit (CLU), the polarization modulator (PMU), and a tip-tilt mirror (CTM) behind the primary mirror. The PMU is a continuously rotating waveplate optimized for linear and circular polarization at 5,173 and 6,302 Å. The SOT is well designed and calibrated for performing polarization measurements with high accuracy (Ichimoto *et al.*, 2007). With the CLU and the CTM, the OTA provides a pointing-stabilized parallel beam to the FPP. The FPP has four optical paths: the Narrowband Filter Imager (NFI), the Broadband Filter Imager (BFI), the Spectro Polarimeter (SP), and the Correlation Tracker (CT). The BFI and the NFI share a CCD detector and

Table 1 The *Hinode* mission.

Mission objective	Investigation of magnetic activity of the Sun including its generation, energy transfer, and release of magnetic energy
Launch	22 September 2006, 21:36 UTC
Mission life	$\geq$3 years
Organization	
Project manager	T. Kosugi[1] (ISAS/JAXA)
Co-manager	S. Tsuneta (NAOJ)
Project scientists	T. Sakurai (NAOJ), K. Shibata (Kyoto University), J.M. Davis (MSFC), and L.K. Harra (MSSL)[2]
Principal investigators	
Solar Optical Telescope (SOT)	S. Tsuneta (NAOJ) and A.M. Title (LMATC)[3]
EUV Imaging Spectrometer (EIS)	J.L. Culhane (MSSL),[4] G.A. Doschek (NRL), and T. Watanabe (NAOJ)
X-Ray Telescope (XRT)	L. Golub (SAO)[5] and K. Shibasaki (NAOJ)
Responsible institutes	
The Japan Aerospace Exploration Agency's Institute of Space and Astronautical Science (ISAS/JAXA)	Overall mission including the launch vehicle
National Astronomical Observatory of Japan (NAOJ)	Three scientific instruments and support for spacecraft development
The National Aeronautics and Space Administration (NASA)	Three scientific instruments
The Particle Physics and Astronomy Research Council (PPARC)[6]	EIS
European Space Agency (ESA)	Ground station support
Major participating institutions	
Scientific instruments are built by collaborative efforts of the following institutes	
SOT	NAOJ, Lockheed Martin Solar and Astrophysics Laboratory (LMSAL), High Altitude Observatory (HAO), ISAS/JAXA, NASA
EIS	Mullard Space Science Lab. (MSSL), US Naval Research Laboratory (NRL), NAOJ, ISAS/JAXA, Rutherford Appleton Laboratory (RAL), Birmingham University, The University of Oslo
XRT	Smithsonian Astrophysical Observatory (SAO), ISAS/JAXA, NAOJ, NASA

[1] I. Nakatani as project manager and T. Sakao and T. Shimizu as deputy project managers after T. Kosugi passed away in November 2006.

[2] Succeeded by D.R. William in 2006.

[3] Succeeded by T.D. Tarbell in 2004.

[4] Succeeded by L.K. Harra in 2006.

[5] Succeeded by E.E. Deluca in 2005.

[6] Now the Science and Technology Facilities Council (STFC).

Table 2 *Hinode* scientific instruments.

(a) Properties of the telescopes

Solar Optical Telescope (SOT)

 Optical Telescope Assembly (OTA)

Optics	Aplanatic Gregorian with aperture of 50 cm

 Focal Plane Package (FPP)

Wavelength and lines	Broadband Filter Instrument (BFI)
	CN (3883.0), Ca II H (3968.5), CH (4305.0)
	Blue (4504.5), Green (5550.5), Red (6684.0)
	Narrowband Filter Instrument (NFI)
	Mg Ib (5172.7), Fe I (5250.2, 5247.1, 5250.6),
	Fe I (5576.1), Na I (5895.9), Fe I (6302.5, 6301.5),
	H I (6562.8)
	Spectro Polarimeter (SP)
	Fe I (6302.5, 6301.5)
Sensitivity to magnetic fields	longitudinal: $1-5$ G
	transverse: $30-50$ G
Typical time cadence	Ranges from tens of seconds for photospheric images and vector magnetographs in particular lines to ≈ 1 hr for the full Stokes profiles

EUV Imaging Spectrometer (EIS)

Optics	Off-axis paraboloid with multilayer-coated mirror and concave grating with aperture of 15 cm
Wavelength	$170-210$ Å with spectral resolution of ≈ 4000
	$250-290$ Å with spectral resolution of ≈ 4600
Velocity resolution	3 km s^{-1} for Doppler velocity, 20 km s^{-1} for line width
Exposure time	Milliseconds in flares, tens of seconds in active regions

X-Ray Telescope (XRT)

Optics	Modified Wolter type I grazing incidence mirror and co-aligned optical telescope
Wavelength	X ray: $2-200$ Å
	Optical: G-band (4305 Å)
Temperature discrimination	$\Delta \mathrm{Log}\, T : 0.2$[1]
Exposure time	4 ms -10 s

(b) Properties of the focal plane detectors

Instruments	F.O.V. EW × NS (slit/slot width)	Pixel size
SOT		
NFI[2]	$328'' \times 164''$	$0.08''$
BFI[2]	$218'' \times 109''$	$0.053''$
SP	$320'' \times 164''\,(0.16'')$	$0.16'' \times 21.5$ mÅ
EIS	$590'' \times 512''\,(1'', 2'', 40'', 266'')$	$1.0'' \times 0.0223$ Å
XRT	$2048'' \times 2048''$	$1.0''$

[1] In the case of isothermal plasma.

[2] NFI and BFI share a CCD.

constitute the Filtergraph (FG). The SP and the CT have their own CCD detectors. The NFI uses a tunable Lyot, birefringent filter to record filtergrams, Dopplergrams, and longitudinal and vector magnetograms across the spectral range from 5170 to 6570 Å. The BFI has interference filters to image the photosphere and low chromosphere and to make blue, green, and red continuum measurements for irradiance studies. The SP is an off-axis Littrow echelle spectrograph that records dual-line, dual-beam polarization spectra of the Fe I 6302.5 Å and 6301.5 Å spectral lines for high-precision Stokes polarimetry. The CT is the high-speed CCD camera to sense jitter of solar features on the focal plane. The jitter signal is fed to the closed-loop control of the tip-tilt mirror (Shimizu *et al.*, 2007). This image-stabilization system prevents the spacecraft jitter from affecting the resolution of the images. The image-stabilization system achieves a stability of $0.007''$ (3σ) below the cross-over frequency of 14 Hz. Time-line sequence of the data acquisitions by the SOT is controlled according to two observation tables (one for FG and the other for SP) on the Mission Data Processor (MDP).

2.2. EUV Imaging Spectrometer

The EIS (Culhane *et al.*, 2007) is an imaging spectrometer designed to observe plasmas in the temperature range from 0.1 MK, the upper transition, to 10 MK, the lower corona. The EIS is an off-axis paraboloid telescope with a focal length of 1.9 m and a mirror diameter of 15 cm. The angular resolution of the optics is $2''$. The total length of the instrument is 3 m. The primary mirror has a mechanism that can offset the field of view of the EIS in the E–W direction relative to the spacecraft pointing. The mirror illuminates various slits that are placed at the focus of two multilayer-coated, toroidal gratings that disperse the spectrum onto two back-side-illuminated CCD detectors. The detectors cover the wavelength ranges of 170–210 Å and 250–290 Å with spectral resolution of $R \approx 4000$. Four slit or slot widths are available: $1''$ slit, $2''$ slit, $40''$ slot, and $266''$ slot. High-spectral-resolution images can be obtained by rostering with the slit. The slots provide "overlappograms" of the transition region and corona at high cadence. The EIS instrument provides a factor of 3 improvement in spatial and spectral resolution and sensitivity over the CDS (Coronal Diagnostic Spectrometer) aboard the SOHO (*Solar and Heliospheric Observatory*) spacecraft. The velocity resolution is 3 km s^{-1} for Doppler velocities and 20 km s^{-1} for line widths. With the higher sensitivity and higher telemetry rate of the spacecraft, the EIS can achieve a time cadence of 0.5 s in flares and $\approx$10 s in active regions. The control system is designed to optimize the use of the telemetry allocation. It provides the flexibility to select the mix of spectral lines, image regions, and time cadence of an observation to match specific scientific objectives. A dedicated processor within EIS provides the control function and can operate autonomously to switch observations in response to notification of a flare by the XRT or detection of a flare or a bright point by the EIS processor itself.

2.3. X-Ray Telescope

The XRT is a grazing incidence telescope of a Wolter I design made from Zerodur (Golub *et al.*, 2007). The mirror has a 30-cm aperture and a 2.7-m focal length. The surface figure is a modified paraboloid-hyperboloid whose surfaces are optimized to minimize the blur circle radius at large angles. The reflecting surfaces are uncoated and, together with improved entrance filters that reject the Sun's visible light, provide a lower energy X-ray cutoff than SXT aboard *Yohkoh*. In front of the focal plane, there are two filter wheels containing a total of nine X-ray analysis filters, which pass wavelength bands with different lower cutoff energy. Because of the lower cutoff energy, the XRT can observe plasmas with temperatures as low as 1×10^6 K in the lower corona. The brightness ratio between images taken through two different filters provides a measure of the temperature of the plasma when the observed plasma can be assumed to be isothermal. For flare studies the filter ratio method is capable of measuring temperatures as high as $\approx 30 \times 10^6$ K. In addition to the X-ray optics, the XRT is equipped with visible light optics, to be used with a G-band filter, for the purpose of co-alignment of XRT and SOT images. The X-ray and visible light optics share the focal plane where a back-side-illuminated CCD is located. The CCD has a pixel size of 1 arcsec and the field of view is 34×34 arcmin^2, which covers the whole solar disk when the spacecraft is pointed at sun center. The CCD camera is equipped with an on-orbit focus adjustment mechanism (Kano *et al.*, 2007). The camera is launched out of focus and in addition to moving the camera to the best on-orbit focus it can also be used to optimize across the field of view to compensate for field curvature. For example, for the highest resolution observations an on-axis focus provides an angular resolution of ≈ 1 arcsec within a radius of ≈ 7 arcmin. For the best resolution across the field of view the focus can be moved forward to provide an angular resolution of ≤ 3 arcsec within a radius of ≈ 17 arcmin. The camera, its shutter for exposure control, and the filter wheel are controlled according to an observation sequence defined as the observation table in the MDP. To optimize the use of the telemetry allocation the field of view, filter sequence, and time cadence can be adjusted to match each scientific objective. MDP also has various functions for enhancing XRT observations, including automatic region selection, automatic exposure duration control, flare detection, and memory buffer for storing high-cadence images taken in the pre-flare phase.

3. The Spacecraft

3.1. General

The *Hinode* spacecraft was launched from the Uchinoura Space Center, located at latitude 31 N, longitude 131 E, by the seventh, and last, M-V launch vehicle into an elliptical polar orbit with a perigee of ≈ 280 km and apogee of ≈ 686 km. In the succeeding phase, the *Hinode* spacecraft boosts its perigee and controls the plane of the orbit with its own thrusters to acquire a circular, sun-synchronous, polar orbit of about 680-km attitude, 98.1-deg inclination, and 98-min period. With this orbit, *Hinode* can observe the Sun continuously for a duration of nine months each year.

The major parameters of the spacecraft are summarized in Table 3. The spacecraft, schematically shown in Figure 1, has dimensions of approximately $4000 \times 1600 \times 1600$ mm with two external solar panels (4300×1100 mm each) and weights about 900 kg. Three telescopes are aligned in the Z-axis of the spacecraft and supported by an optical bench unit (OBU). The OBU is a cylinder made up of composite material that supports the OTA internally. The FPP, EIS, and XRT are kinematically mounted on the outside of the OBU with six mounting legs, which constrain the degrees of freedom of the rigid body. The OBU also holds a tower to whose upper surface the sun sensors are attached. The electronics units are located in the bus box attached to the bottom of the OBU. The solar cell panels are designed to supply about 1100 W during each spacecraft day. Excess power is either stored in NiCd batteries to supply the power required during spacecraft night or is consumed by a shunt regulator.

Table 3 Major parameters of *Hinode*.

Size	4000 × 1600 × 1600 mm
Weight	900 kg (wet), 770 kg (dry)
Power	1100 W
Data rate	Up to 2 Mbps (science data), and 32 kbps (housekeeping)
Data recorder	8 Gbits
Telemetry rate	32 kbps (S-band), 4 Mbps (X-band)
Orbit	
Altitude	680 km (circular, Sun-synchronous, polar orbit)
Inclination	98.1 deg
Period	98 min
Attitude control (requirement)	Three-axis stabilized
Absolute pointing	$20''$
Stability around	X/Y-axes: $0.06''$ (>20 Hz), $0.6''/2$ s, $4.5''/1$ hr
	Z-axis: $200''/1$ hr
Pointing determination	X/Y-axes: $0.1''$
Offset pointing	Up to $1178''$ from the Sun center
Ground stations	
Commanding and downlink	Uchinoura Space Center (131 E, 31 N)
Commanding only	JAXA new Ground Network stations
Downlink only	Svalbard (15 E, 78 N)
Number of downlinks	15 per day (Svalbard)
	4 per day (Uchinoura)

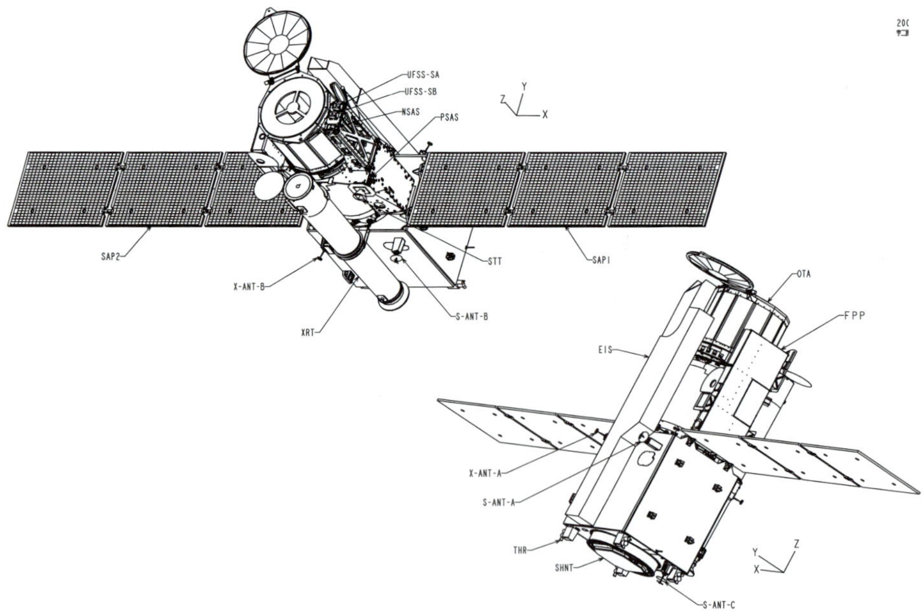

Figure 1 The *Hinode* spacecraft and its scientific instruments.

3.2. Attitude and Orbit Control

The *Hinode* spacecraft is stabilized by the attitude and orbit control system (AOCS) in three axes with its Z-axis pointed to the Sun. The Y-axis is directed toward solar north. As a baseline, the spacecraft tracks a region on the solar surface by correcting for solar rotation. For each tracked target, the angular velocity around the rotation axis of the Sun can be specified. The other mode is the spacecraft pointing to a fixed position on the solar disk. In either case, stability of the Z-axis is $0.3''$ (3σ) per 10 s and $1''$ per min.

The AOCS uses momentum wheels and magnetic torquers as the actuators for attitude control and thrusters for orbital control. The attitude sensors, including two fine sun sensors (UFSS), a star tracker, and geomagnetic sensors are available for determining spacecraft pointing relative to the direction of the Sun and to the ecliptic plane while an inertial reference unit comprising four gyros detects changes of attitude with time. Signals from two UFSS sun sensors with random noise level of $0.3''$ (3σ) can be used to remove the jitter of the satellite Z-axis pointing from the time series of data.

3.3. Command System

An uplink commanding system controls the operation of all the instruments on the spacecraft. Commands are sent from the Uchinoura Space Center as well as from JAXA new Ground Network antennas. There are about three contacts in a day for commanding. Each contact has a duration of up to 10 min. Commands from the ground are received by the command unit and distributed by the data handling unit (DHU). Commands for the scientific instruments are further relayed by the MDP. The DHU can coordinate commands into sequences called organized commands (OGs). The DHU can store up to 512 sets of OGs, each being a set of up to eight commands. First, an OG can be launched by a "real-time OG execute command" from the ground. Second, a series of OGs can be dispatched sequentially with specified time intervals by the DHU itself. Such a series is called an operation program (OP). The OP can contain up to 4096 OG references. The OP is initiated by an "OP start command." The OP can last for up to about 10 days, so that the operation can be programmed beforehand. In addition to the OG, the DHU can store sequences of commands to be executed during spacecraft emergencies. These are triggered by the AOCS or by an autonomous detection of an emergency by the DHU. The latter case includes failure modes of the battery system.

3.4. Onboard Data Processing

Observations of the three scientific instruments are governed by the MDP. Figure 2 is a schematic representation of the onboard observation control system. In the case of the FG, SP, and XRT, the MDP controls the observations. The controls are implemented using observation tables that make use of programs that have a nested loop – call structure. The EIS instrument's observing sequences are controlled by its own processor. In addition to normal observations, the scientific instruments have the capability to switch to autonomous observations when notified by the onboard system of a flare. The MDP continuously analyzes XRT images for large intensity increases indicative of a flare. If a flare is found a flare flag is issued that allows the instruments to terminate their current sequence and switch to a flare observation program. The observation table for flare studies has the same structure as those for normal observations.

The scientific data from the instruments are compressed in the MDP before being stored in the data recorder. Memory space is divided among the SOT, the XRT, and the EIS in

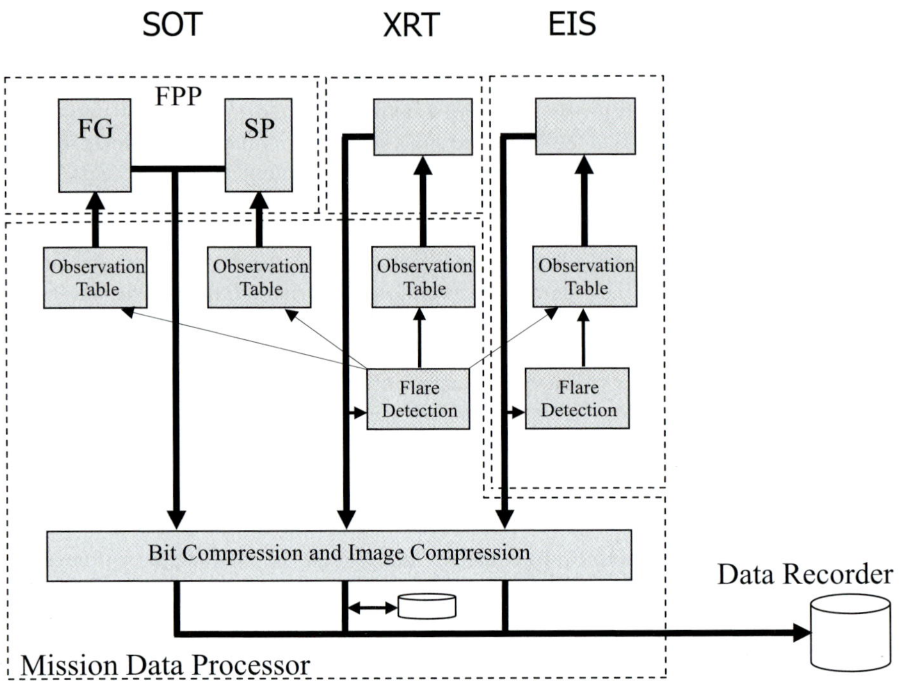

Figure 2 Functional block diagram of onboard observation control system.

the ratio of 70 : 15 : 15 for many periods of observations. The MDP has a compression speed of 832 kpixel s^{-1} for SOT, 256 kpixel s^{-1} for XRT, and 128 kpixel s^{-1} for EIS, which matches the data acquisition rate and storage capacity for each instrument. Two types of compression are performed sequentially. The first is pixel-by-pixel bit compression followed by image compression. The pixel-by-pixel bit compression is based on look-up tables and implemented by hardware. In the table a smooth function composed of linear and quadratic components can be registered. The image compression is either a lossless compression using a DPCM (Differential Pulse Code Modulation) algorithm or a JPEG (Joint Photographic Experts Group) lossy compression using a DCT (Discrete Cosine Transform) algorithm. These schemes are implemented by an application-specific integrated circuit (ASIC). Parameters for compressions, which affect the compression ratio and data quality, can be optimized on orbit.

The MDP can output compressed data from the SOT, XRT, and EIS at rates of up to 1.3 Mbps, 262 kbps, and 262 kbps, respectively. The actual data rates from the telescopes are determined by the observation tables and compression efficiency. During the preparation of the observation table care has to be taken to ensure that they are consistent with the duration of downlink contacts. The tables should be implemented in scientific operation as described in Section 4.2.

3.5. Data Recording

Telemetry from the spacecraft follows the data packet specification recommended by the Consultative Committee for Space Data Systems (CCSDS). Telemetry packets from the scientific instruments are edited by the MDP; the housekeeping and spacecraft data are edited by the DHU.

The *Hinode* spacecraft has limited-duration ground station contacts. Telemetry packets that cannot be downlinked during a particular ground station contact remain stored in the onboard data reorder and are played back in following contacts. The data recorder has two partitions. Data from the spacecraft and scientific instruments are stored in separate partitions, so that scientific operations do not conflict with maintaining the integrity of the spacecraft data. Each partition has a write pointer for recording and a read pointer for playback and behaves like a first-in-first-out (FIFO) memory. When a partition becomes full, the data recorder either overwrites the oldest data or stops recoding according to its setting.

The recorder memory has a total capacity of 8 Gbits. This capacity is three times greater than the amount of data that can be downlinked during a ground station contact. Distribution of ground contacts in a day can be irregular. With the large capacity of the data recorder, the data rate from telescopes can be determined on a daily basis rather than from the distribution of ground contacts. This feature is well suited for continuous observation in the sun-synchronous orbit of *Hinode*. The three telescopes share their partition of the data recorder. Unexpected data volume from one telescope (*e.g.* from human error in observation planning or degraded compression efficiency) can result in a loss of data for the other telescopes. To prevent this from happening the MDP can be programmed to prevent any of the three telescopes from exceeding its allocation. The MDP monitors the cumulative data recorded by each telescope until a specified limit is reached, at which time it stops further packet addition for that telescope. The assignment of data among the telescopes can be changed on a daily basis.

3.6. Telemetry

Data acquired with the instruments onboard *Hinode* are downlinked to Uchinoura Space Center station as well as the Norwegian high-latitude ($78°14'$ N) ground station at Svalbard. Svalbard downlinks for every station contact are realized by cooperation between the European Space Agency (ESA) and the Norwegian Space Centre. Two telemetry channels, S-band (2.2 GHz) and X-band (8.4 GHz), are used. The S-band channel transmits real-time status at 32 kbps. The X-band transmits all of the real-time data and recorded data from the data recorder at 4 Mbps. At the Uchinoura station, the two channels are received simultaneously. At the Svalbard station only recorded data are transmitted via the X-band and no real-time data are available. Note that only real-time data are transmitted via the S-band at the JAXA Ground Network stations for commanding purposes. During the downlink, real-time transfer has higher priority than that of recorded data from the data recorder. Real-time data downlinked at Uchinoura are sent to ISAS at Sagamihara, near Tokyo, with the Space Data Transfer Protocol (SDTP) over a TCP/IP network. Recorded data are also sent to ISAS within 90 min of the downlink. Data taken at the Svalbard station are transmitted to ISAS through the Internet, nominally within 90 min of their receipt, where they are combined with the data from the Uchinoura station and placed into the ISAS Sirius database.

From the Sirius database, the data are reformatted into FITS files and classified as Level 0 data and archived on the ISAS DARTS system from where they are made available to the scientific community. The master archive is mirrored to the Solar Data Analysis Center (SDAC) at the Goddard Space Flight Center in Greenbelt, Maryland, and also to data centers in Norway and at MSSL. The principal investigator institutions in Europe and the United States and several co-investigator institutions mirror the data from their instrument to their home institutions.

4. Operations

4.1. Initial Operations and Observations

The month following launch is a period for checking out the spacecraft and the instruments where the spacecraft and the instruments are operated by their builders. After this period, observation with the scientific instruments starts. The first 90 days of observations are planned before the launch and are conducted by the *Hinode* Principal Investigator teams. This provides the instrument teams with the opportunity to learn the operational skills needed to run the mission, including the scheduling of operations and archiving data. During this period there are occasional opportunities to access the data archive to retrieve specific data sets. These opportunities enable the user community to test the system and help identify problems before the full data set is released. This is planned to occur about six months after initial operations begin. At that time all the archived data shall be available and all new observations shall be released as soon as after their acquisition.

4.2. Spacecraft Operation

The *Hinode* orbit provides at least two morning and two evening contacts in Japan. Morning contacts provide quick-look science data and the evening contacts are used for uploading commands to the spacecraft and science instruments. In addition to the Japanese contacts, the ESA provides 15 contacts per day through Svalbard for downloading scientific data. The average contact time at Svalbard is 11.5 min. By allowing 15 s for handshaking, approximately 42.5 Gbits of data are downloaded per day.

After the initial period, it is expected that the operation of the spacecraft will become routine. To facilitate safe operation of the spacecraft, patterns of the operations are accumulated and maintained in a knowledge base. In daily operations, a planning tool generates commands for the spacecraft using the knowledge base, predictions of orbital conditions, and specification of the downlink stations. The tool also calculates the telemetry allocation for the scientific instruments to be used in planning the scientific program and merges the spacecraft and scientific operations.

4.3. Scientific Operation

Scientific operations are conducted from the ISAS facility located in Sagamihara, Japan. They are separated into planning and implementation. As shown in Figure 3, the planning process involves monthly, weekly, and daily planning meetings. Monthly meetings or teleconferences establish the high-level objectives for the next three months and more detailed objectives for the next month. The goal of these meetings is to approve and schedule observing proposals from the external community that were submitted to and approved by the Scientific Schedule Coordinators (SSC). The SSCs are senior scientists designated by the instrument Principal Investigators (PIs) who reside at their home institutions. They are responsible for coordinating the monthly observation schedules proposed by the instrument teams with the external proposals. They are also available to assist the external community in preparing proposals and identifying contacts within the instrument teams who can provide proposers with the detailed capabilities of their instruments.

Weekly meetings are held each Friday at ISAS and establish the observing plan, subject to minor changes, for the next week. The plan includes target regions, pointing maneuvers, and data recorder allocations. The plan is placed on the *Hinode* operation Web sites to allow

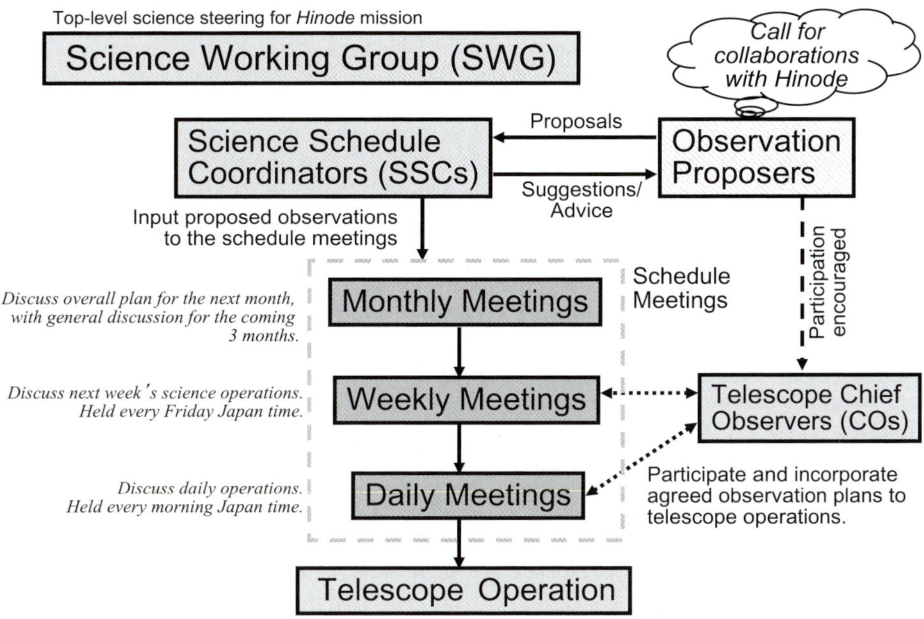

Figure 3   Scientific operation planning flow of *Hinode*.

coordination with other observatories. The daily meetings are held six mornings a week at ISAS at 10:30 AM local time (01:30 UT), during which the daily plan is finalized. In the planning context "days" start at the spacecraft's evening contacts in Japan, which occur at approximately 4:00 – 7:30 PM local time or 7:00 – 10:30 UT. At these contacts the instrument commands and observing tables for the next 24 hours are uplinked to the spacecraft. With this planning schedule it is possible, in principle, to make minor adjustments to the observing plan as little as eight hours before the observations are made.

4.4. Community Involvement

The *Hinode* science teams hope and expect that *Hinode* proves to be a valuable asset to the international scientific community. To expedite collaboration we have created the role of Scientific Schedule Coordinator to provide an interface to the experiment teams and to educate proposers as well as review proposals and schedule observations. Collaboration with other observatories, missions, campaigns, or suborbital programs are given high priority. However, the data from these observations are also freely available to the community (Matsuzaki *et al.*, 2007).

5. Concluding Remarks

Hinode is a complex satellite that is designed to study primarily how changes in the magnetic field as it emerges through the photosphere affect the higher levels in the atmosphere. It is hoped that the high-resolution observations of the vector magnetic field clarify the conditions needed for the onset of magnetic reconnection. The development of the science instruments and objective has been and remains a truly international program and it is hoped that an even broader group of the world's scientists participate in the observations and their analysis.

Acknowledgements   The authors would like to acknowledge the people who contributed to the spacecraft design, development, and tests. We express our gratitude to those who were involved in the spacecraft preparation:

ISAS/JAXA: N. Bando, E. Hirokawa, K. Hirose, T. Ichikawa, K. Inoue, N. Ishii, T. Kato, J. Kawaguchi,Y. Maeda, Y. Mochihara, O. Mori, Y. Morita, T. Nagae, H. Nakabe, J. Nakatsuka, H. Saito, S. Sakai, H. Sawai, M. Shida, T. Shimada, K. Shimomura, S. Shimose, K. Shuto, T. Takemae, H. Takeuchi, M. Tamura, M. Tajima, T. Toda, H. Toyoata, H. Yamakawa, T. Yamamoto, Z. Yamamoto, Y. Yoshida, and M. Yoshikawa.

NAOJ: R. Kano, Y. Katsukawa, M. Nakagiri, T. Tamura, and T. Bando.

Mitsubishi Electric Co. (MELCO): J. Akiyama, Y. Aoki, T. Hashizume, K. Hiraide, T. Hayashi, T. Inoue, Y. Ito, H. Izu, T. Kamachi, M. Kasama, K. Kidoguchi, M. Koike, T. Kosuge, K. Nakagawa, M. Mitsutake, T. Sato, Y. Shirahama, T. Shiraishi, K. Takeo, H. Tomoeda, and M. Yoshimura.

NEC Toshiba Space Systems (NTS): T. Abe, K. Fujiwara, T. Gondai, Y. Haruna, I. Kanaoka, N. Kaneko, M. Kubo, T. Kumai, Y. Okada, T. Osashima, M. Matsui, S. Murata, T. Okumura, N. Ogura, T. Saito, T. Shimamura, K. Taniguchi, K. Tsuno, S. Tsuruta, and H. Yamaki.

Furukawa Battery: H. Inafuku.

Mitsubishi Heavy Industries (MHI): K. Furukawa, K. Hisatsune, M. Koyama, and T. Takami.

Panasonic System Solutions: T. Watanabe, G. Furuhashi, and K. Nemoto.

Japan Aviation Electronics Industry (JAE): K. Hattori and S. Miyahara.

Fujitsu: Y. Iizuka, M. Kojima, T. Kosaka, M. Morita, S. Nagata, M. Yokoyama, and M. Yamashita.

GN/JAXA: T. Fuse, S. Hirose, K. Narita, and T. Saito.

We should note that many more engineers, technicians, scientists, and administrators made their contributions to the *Hinode* project. The authors also express their thanks to the M-V rocket team led by Y. Morita for successfully installing the spacecraft into the orbit.

References

Acton, L., Tsuneta, S., Ogawara, Y., Bently, R., Bruner, M., Canfield, R., *et al.*: 1992, *Science* **258**, 618.
Canfield, R.C., Hudson, H.S., McKenzie, D.E.: 1999, *Geophys. Res. Lett.* **26**, 627.
Culhane, J.L., Harra, L.K., James, A.M., Al-Janabi, K., Bradley, L.J., Chaudry, R.A., *et al.*: 2007, *Solar Phys.*, in press.
Golub, L., DeLuca, E., Austin, G., Bookbinder, J., Caldwell, D., Cheimets, P., *et al.*: 2007, *Solar Phys.*, in press.
Harra, L.K., Schmieder, B., van Driel-Gesztelyi, L., Sato, J., Plunkett, S.P., Rudawy, P., Rompolt, B., Akioka, M., Sakao, T., Ichimoto, K.: 1998, *Astron. Astrophys.* **337**, 911.
Harra, L.K., Matthews, S.A., van Driel-Gesztelyi, L.: 2003, *Astrophys. J.* **598**, L59.
Hudson, H.S., Khan, J.I., Lemen, J.R., Nitta, N.V., Uchida, U.: 2003, *Solar Phys.* **212**, 121.
Ichimoto, K., Lites, B., Elmore, D., Suematsu, Y., Tsunete, S., Katsukawa, Y., *et al.*: 2007, *Solar Phys.*, submitted.
Kano, R., Sakao, T., Hara, H., Tsuneta, S., Matsuzaki, K., Kumagai, K., *et al.*: 2007, *Solar Phys.*, submitted.
Khan, J.I., Hudson, H.S.: 2000, *Geophys. Res. Lett.* **27**, 1083.
Masuda, S., Kosugi, T., Hara, H., Tsuneta, S., Ogawara, Y.: 1994, *Nature* **371**, 495.
Matsuzaki, K., Shimojo, M., Tarbell, T.D., Harra, L.K., DeLuca, E.: 2007, *Solar Phys.*, in press.
Ogawara, Y., Takano, T., Kato, T., Kosugi, T., Tsuneta, S., Watanabe, T., Kondo, I., Uchida, Y.: 1991, *Solar Phys.* **136**, 1.
Priest, E., Parnell, C., Martin, S.F.: 1994, *Solar Phys.* **427**, 459.
Shibata, K., Masuda, S., Shimojo, M., Hara, H., Yokoyama, T., Tsuneta, S., Kosugi, T., Ogawara, Y.: 1995, *Astrophys. J.* **451**, L83.
Shimizu, T.: 1995, *Publ. Astron. Soc. Japan* **47**, 251.

Shimizu, T., Shine, R.A., Title, A.M., Tarbell, T.D., Frank, Z.: 2002, *Astrophys. J.* **574**, 1074.

Shimizu, T., Nagata, S., Tsuneta, S., Tarbell, T., Edwards, C., Shine, R., *et al.*: 2007, *Solar Phys.*, submitted.

Shimojo, M., Hashimoto, S., Shibata, K., Hirayama, T., Hudson, H.S., Action, L.W.: 1996, *Publ. Astron. Soc. Japan* **48**, 123.

Sterling, A.C., Hudson, H.S., Thompson, B.J., Zarro, D.M.: 2000, *Astrophys. J.* **532**, 628.

Suematsu, Y., Tsuneta, S., Ichimoto, K., Shimizu, T., Otsubo, M., Katsukawa, Y., *et al.*: 2007, *Solar Phys.*, submitted.

Tarbell, T.D., *et al.*: 2007, *Solar Phys.*, submitted.

Tsuneta, S.: 1996, *Astrophys. J.* **456**, 840.

Tsuneta, S., Suematsu, Y., Ichimoto, K., Shimizu, T., Otsubo, M., Nagata, S., *et al.*: 2007, *Solar Phys.*, submitted.

Data Archive of the Hinode Mission

**K. Matsuzaki · M. Shimojo · T.D. Tarbell · L.K. Harra ·
E.E. Deluca**

Originally published in the journal Solar Physics, Volume 243, No 1.
DOI: 10.1007/s11207-006-0303-2 © Springer 2007

Abstract All of the *Hinode* telemetry data are to be reformatted and archived in the DARTS
system at ISAS and mirrored to data centers around the word. The archived data are distributed to users through the Internet. This paper gives an overview of the files in the archive,
including the file formats. All formats are portable and have heritage from the previous missions. From the reformatted files, index information is created for faster data search. Users
can perform queries based on information contained in the index. This allows for searches
to return observations that conform to particular observing conditions.

1. Introduction

In solar physics, coordinated analyses of observational data from different instruments are
crucial to reveal the nature of phenomena. *Hinode*, which had been called Solar B before
its launch, is an observatory-style mission equipped with instruments for coordinated observation of the solar atmosphere at different altitudes and aims at revealing how magnetic

K. Matsuzaki (✉)
Institute of Space and Astronautical Science, Japan Aerospace Exploration Agency, Sagamihara,
229-8510 Kanagawa, Japan
e-mail: matuzaki@solar.isas.ac.jp

M. Shimojo
Nobeyama Solar Radio Observatory, National Astronomical Observatory of Japan, Nobeyama,
384-1305 Nagano, Japan

T.D. Tarbell
Lockheed Martin Solar and Astrophysics Laboratory, B/252, 3251 Hanover Street,
Palo Alto, CA 94204, USA

L.K. Harra
UCL Mullard Space Science Laboratory, Holmbury St. Mary, Dorking, RH5 6NT Surrey, UK

E.E. Deluca
Smithsonian Astrophysical Observatory, Cambridge, MA 02138, USA

energy is transferred from the photosphere to the atmosphere, resulting in energy release (Kosugi *et al.*, 2006). There are three instruments aboard *Hinode*: the Solar Optical Telescope (SOT), which observes the photosphere and the chromosphere, the EUV Imaging Spectrometer (EIS), which observes the transition region, and the X-ray Telescope (XRT), which observes the corona. The instruments mainly observe the same target at which the spacecraft is pointed. Once data are obtained from the instruments, combined analyses of the data are of great importance. To maximize the scientific analysis, an archive of the *Hinode* mission data is maintained. This paper describes how the archive is designed, developed, and operated.

The data archive of *Hinode* is built on the heritage of previous solar missions. For *Yohkoh*, whose success motivated the *Hinode* mission, a common reformatted database for all instruments (Morrison *et al.*, 1991) was created. The data archive of *Hinode* inherits its basic design of data flow from the *Yohkoh* reformatted database. The *Yohkoh* reformatted database was initially distributed with off-line media as 8-mm tapes. The *Yohkoh* mission continued for nearly a solar cycle. During this time, the Internet spread over the world and the Institute of Space and Astronautical Science (ISAS) started service of the Data ARchive and Transmission System (DARTS; Miura *et al.*, 2000), which provides the science data of the ISAS missions, including the contents of the *Yohkoh* reformat database, over the Internet.

The *Yohkoh* reformatted database has been supplied with a software package for data analysis. The package has been written in Interactive Data Language (IDL) of Research Systems, Inc. The combination of the package and programming environment of the IDL is suitable for manipulation and visualization of data by researchers of solar physics. Thus, the software package evolved into a common programming and data analysis environment covering numerous solar physical missions and is named SolarSoft (Freeland and Handy, 1998). Following the framework of the previous missions, the data archive of *Hinode* is to be supplied with a software package of the data analysis integrated into SolarSoft. In this paper, we describe the data archive of *Hinode* from the viewpoints of data flow in Section 2, the contents in Section 3, and the user interface in Section 4.

2. Data Flow

Data flow of the *Hinode* mission is shown in Figure 1. The *Hinode* spacecraft and its scientific instruments generate telemetry packets in the format defined by the recommendation of the Consultative Committee for Space Data Systems (CCSDS). The telemetry is received by the Uchinoura Space Center (USC) station in Japan and by the Svalbard station near the North Pole. The spacecraft uses a data recorder (DR) to store the data when a real-time station downlink is not available. The 8 Gbit capacity of the DR can store more than three times the amount of data that can be downlinked in a ground station contact. There are 3 to 4 downlinks to the USC stations and 15 downlinks to the Svalbard stations in a day. The USC data reach ISAS within two hours from the down link, while the transfer of the Svalbard data may take a day. All of the *Hinode* raw telemetry data are time-ordered and stored on-line in the SIRIUS database at ISAS. The maximum data downlinked is 160 Gbyte per month. The SIRIUS database sends raw telemetry of CCSDS packets into workstations at ISAS, which generate reformatted data files. The initial reformatting is performed for the USC data within a few hours from the downlink. Reformatting covering both the USC data and the Svalbard data is performed half a day after and also a few days after the downlink. The files generated are stored in the archive system DARTS (Tamura *et al.*, 2004) at ISAS. The master archive of *Hinode* in DARTS is mirrored into data centers around the world over the Internet. From

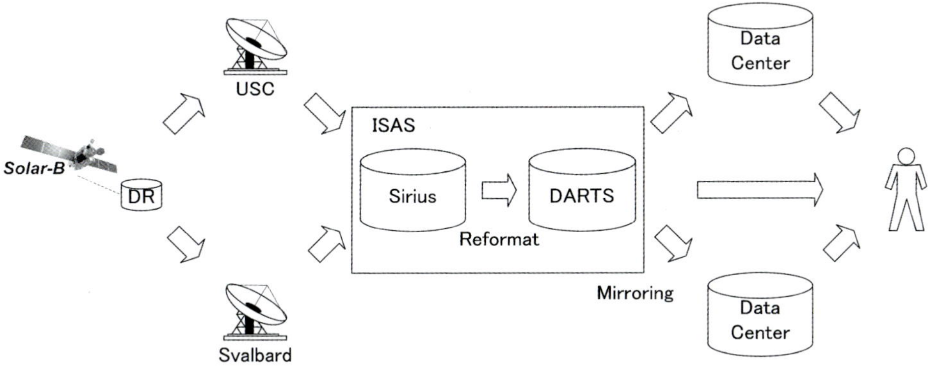

Figure 1 *Hinode* data flow.

DARTS and mirrored data centers, data are distributed to users through the Internet. In the next section, we describe the contents of the archive, including the reformatted data files.

In addition to the telemetry data, an observation database that includes the observation plans and observation logs is also accumulated by each instrument's team. Basically, these data are created on the ground independently of the mission archive.

3. The Contents of the Archive

3.1. Reformatted Data Files

All the telemetry from the *Hinode* spacecraft, including that from the scientific instruments, is stored in the computer-readable files of the archive. We intend the files to be portable, readable, and analyzable on any machine. Thus, the telemetry of CCSDS packets from the spacecraft is reformatted into suitable format. The files in the archive are categorized in the following types:

- reformatted science data (Level 0/1),
- reformatted status data,
- calibration data, and
- higher level products (Level 2/Q).

We selected the format of each type so that we can maximize re-use of software developed for instruments in previous missions. In the remainder of this section, we give a brief description of each file type.

Reformatted science data (Level 0/1) are created from the observation data packets from the scientific instruments and from additional information in the observation database. Data files are separately created for SOT/FG, SOT/SP, XRT, and EIS. Here, the SOT files are separated for the FG and the SP, two detectors at the focal plane.

These are further classified into two processing levels. Level 0 data are reformatted raw telemetry data. Level 1 data are fully calibrated data. Raw data used for calibration (*e.g.*, dark images) are included in the Level 0 data, when appropriate. In any case, the reformatted science data has the format of the FITS standard (Hanisch *et al.*, 2001) with binary table extensions (Cotton, Tody, and Pence, 1995) and has a header area. The header is constructed from the telemetry data (*e.g.*, observing time, exposure time, pixel size, coordinates on the

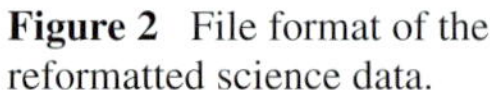

Figure 2 File format of the reformatted science data.

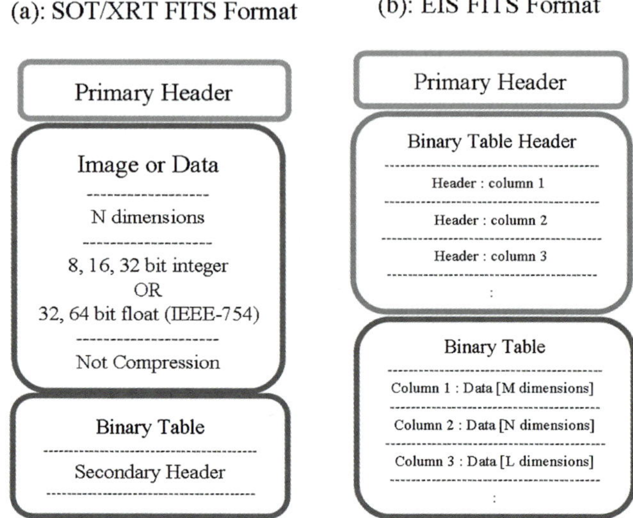

CCD, *etc.*) and the operation database (Table ID, Table name, Comments from Chief Observer, *etc.*). Attitude information included in the Level 0 data is calculated by the attitude and orbit control system (AOCS) aboard the spacecraft and expected to have an accuracy of a few arc sec. The header of the Level 1 data consists of the header of the Level 0 data and additional information from the calibration data (the coordinates on the Sun, *etc.*).

There are two variations in the format of the reformatted science data. For the SOT and XRT, the format shown in Figure 2(a) is applied. In this format, uncompressed image data are written as standard, multi-image FITS files. For the SOT, the unit of the file is the raw CCD image. For the SOT/FG, a file consists of one observable: a single filter image or a set of images obtained to create a single Dopplergram or magnetogram. For the SOT/SP, a file consists of a data cube in four-dimensional space (north – south direction, wavelength direction, four polarization parameters, and two optical paths) at a slit position. For the XRT, a file consists of a single image. For all of the SOT and XRT FITS files, the header information of exposure from scientific instruments is put into the primary HDU. For the EIS, the format shown in Figure 2(b) is applied. This format is similar with that used in SOHO/CDS (St. Cyr *et al.*, 1991). In this format, uncompressed data are put into a binary table extension. The header information, including those for each wavelength, are put in the header of the binary table extension. The unit of the file is one raster scan.

Reformatted status data are created from status data packets from all instruments aboard the spacecraft. File formats of status of the scientific instruments are chosen by each instrument. For data from the spacecraft, FITS formatting with binary table extensions is applied. These files are separately created for each type of CCSDS packet. These files are reformatted for possible use in the future when necessary information in this section of telemetry is not included in the respective scientific data files. There are files that include spacecraft attitude data. In this type of file, spacecraft attitude data calculated from the telemetry of the AOCS is included. This type of file will be updated when knowledge of attitude determination is improved in the future.

Calibration data are data used to derive the Level 1 data from the Level 0 data. The calibration data are made from any available data, including pre-launch test data, the pointing data of the spacecraft, some of the Level 0 data (*e.g.*, dark images), and appropriate spectrum synthesis models. The format of the calibration data is chosen for each datum, depending on the objectives.

Higher level products (Level 2/Q) are created from the reformatted science data (Level 0/1) for various purposes. The Level Q data are created from the Level 0 data with the aim of facilitating reporting and qualitative analysis. An example of the Level Q data is the snapshot movies observed with the XRT. The Level 2 data are created from the Level 1 data and aim at quantitative analysis. An example of the Level 2 data is vector magnetograms created from the SOT/SP Level 1 with inversion algorithms. The format of higher level products is chosen for each product, depending on the objectives. Most of the data products in this category are to be defined after the launch.

3.2. Index Information

A combination of the Level 0 data and the calibration data gives a complete data set to be used in the scientific analysis. However, the data set is too large for searching data for detailed analysis of particular phenomena. For faster data search, index information, which gives a brief summary of the observation data, is created. The index information is automatically created from the Level 0 data and has the format of tab-separated values. Each line of the file presents an observable of the SOT/FG, a map of the SOT/SP, an image of the XRT, and a raster scan of the EIS. Information on the spacecraft orbital solutions, the reformatted data file IDs, and conversion coefficients are also provided in this log. Files are created for the observations in each month. Index information for one month is expected to comprise about 200 Mbytes, which is small enough to be mirrored by data centers and individual researchers.

4. User Interface

Users of the *Hinode* archive can access its contents through the DARTS and the mirroring data centers around the world. Users can obtain any data file and the index information through the Internet, either with the ftp or http protocols, except for the Level 1 data. The files on the archive may be compressed using standard lossless compression schemes (*e.g.*, gzip) to save storage volume and time for data transfer. The Level 1 data are excluded from the initial contents of the archive since a small change of the calibration data causes a change of the Level 1 data in large volume. Instead, individual users are responsible for generating the Level 1 data from the Level 0 and calibration data on their own computers by using the software package in SolarSoft. Revisions of the calibration data are announced via the Internet. Both the *Hinode* package in SolarSoft and the calibration data are distributed over the Internet using the existing distribution scheme of SolarSoft.

For efficient data search by users of the data archive, the index information is registered on relational databases of DARTS and some mirroring data centers. Users can perform queries against the relational database through the Web pages of DARTS and the data centers. For intensive data searches by users against the relational database, a command-line interface working on the user's computer is also provided. In either interface, users can download data determined in the query. Once a user has downloaded data, the user can read the data files with application interfaces included in the *Hinode* package in SolarSoft. Users can perform further analysis either with their own programs or existing analysis tools, such as the package in SolarSoft.

In the near future, it is planned that the data archive of *Hinode* will provide interfaces compatible with the Virtual Solar Observatory (VSO; Hill *et al.*, 2004), European Grid for Solar Observations (EGSO; Bentley, Csillaghy, and Scholl, 2004), or their successors, which improves data accessibility for joint analysis covering different missions.

References

Bentley, R.D., Csillaghy, A., Scholl, I.: 2004, In: Quinn, P.J., Bridger, A. (eds.) *Optimizing Scientific Return for Astronomy through Information Technologies, Proc. SPIE*, vol. **5493**, p. 170.

Cotton, W.D., Tody, D.B., Pence, W.D.: 1995, *Astron. Astrophys. Suppl.* **113**, 159.

Freeland, S.L., Handy, B.N.: 1998, *Solar Phys.* **182**, 497.

Kosugi, T., Matsuzaki, K., Sakao, T., Shimizu, T., Tsuneta, S., Hara, H., Watanabe, T., Davis, J.M., Title, A.M., Culhane, J.L., Harra, L.K., Doscheck, G.A., Golub, L.: 2006, *Solar Phys.*, to be submitted.

Hanisch, R.J., *et al.*: 2001, *Astron. Astrophys.* **376**, 359.

Hill, F., Bogart, R.S., Davey, A., Dimitoglou, G., Gurman, J.B., Hourcle, J.A., Martens, P. C., Suarez-Sola, I., Tian, K., Wampler, S., Yoshimura, K.: 2004, In: Oschmann, J.M., Jr. (ed.) *Ground-Based Telescopes, Proc. SPIE*, vol. **5493**, p. 163.

Miura, A., Shinohara, I., Matsuzaki, K., Nagase, F., Negoro, H., Uno, S., Matsui, S., Watanabe, M., Yamashita, A., Takahashi, H., Matsui, H., Hoshino, M.: 2000, In: Manset, N., Veillet, C., Crabtree, D. (eds.) *Astronomical Data Analysis Software and Systems (ADASS) IX, Astron. Soc. Pac. Conf. Ser.*, vol. **216**, p. 180.

Morrison, M.D., Lemen, J.R., Action, L.W., Bentley, R.D., Kosugi, T., Tsuneta, S., Ogawara, Y., Watanabe, T.: 1991, *Solar Phys.* **136**, 105.

St. Cyr, O.C., Sanchez-Duarte, L., Martens, P.C.H., Gurman, J.B., Larduinat, E.: 1991, *Solar Phys.* **162**, 39.

Tamura, T., Baba, H., Matsuzaki, K., Miura, A., Shinohara, I., Nagase, F., Fukushi, M., Uchida, K.: 2004, In: Ochsenbein, F., Allen, M.G., Egret, D. (eds.) *Astronomical Data Analysis Software and Systems (ADASS) XIII, Astron. Soc. Pac. Conf. Ser.*, vol. **314**, p. 22.

The X-Ray Telescope (XRT) for the *Hinode* Mission

**L. Golub · E. DeLuca · G. Austin · J. Bookbinder · D. Caldwell · P. Cheimets ·
J. Cirtain · M. Cosmo · P. Reid · A. Sette · M. Weber · T. Sakao · R. Kano ·
K. Shibasaki · H. Hara · S. Tsuneta · K. Kumagai · T. Tamura · M. Shimojo ·
J. McCracken · J. Carpenter · H. Haight · R. Siler · E. Wright · J. Tucker ·
H. Rutledge · M. Barbera · G. Peres · S. Varisco**

Originally published in the journal Solar Physics, Volume 243, No 1.
DOI: 10.1007/s11207-007-0182-1 © Springer 2007

Abstract The X-ray Telescope (XRT) of the *Hinode* mission provides an unprecedented combination of spatial and temporal resolution in solar coronal studies. The high sensitivity and broad dynamic range of XRT, coupled with the spacecraft's onboard memory capacity and the planned downlink capability will permit a broad range of coronal studies over an extended period of time, for targets ranging from quiet Sun to X-flares. This paper discusses in detail the design, calibration, and measured performance of the XRT instrument up to the focal plane. The CCD camera and data handling are discussed separately in a companion paper.

L. Golub (✉) · E. DeLuca · G. Austin · J. Bookbinder · D. Caldwell · P. Cheimets · J. Cirtain ·
M. Cosmo · P. Reid · A. Sette · M. Weber
Harvard-Smithsonian Center for Astrophysics, 60 Garden Street, Cambridge, MA 02138, USA
e-mail: lgolub@cfa.harvard.edu

T. Sakao
Institute of Space and Astronautical Science, Japan Aerospace Exploration Agency,
3-1-1 Yoshinodai, Sagamihara, Kanagawa 229-8510, Japan

R. Kano · H. Hara · S. Tsuneta · K. Kumagai · T. Tamura
National Astronomical Observatory, Mitaka, Tokyo 181-8588, Japan

K. Shibasaki · M. Shimojo
Nobeyama Solar Radio Observatory, National Astronomical Observatory, Minamimaki,
Minamisaku, Nagano 384-1305, Japan

J. McCracken · J. Carpenter · H. Haight · R. Siler · E. Wright · J. Tucker · H. Rutledge
NASA/Marshall Space Flight Center, Huntsville, AL 35812, USA

M. Barbera · G. Peres · S. Varisco
Osservatorio Astronomico di Palermo "G.S. Vaiana", Piazza del Parlamento 1, 90134 Palermo, Italy

1. Introduction to XRT and *Hinode*

The solar outer atmosphere presents a unique set of problems for the observer: The temperature varies from 5800 to more than 10^7 K, with a consequent range in primary emission wavelengths; the target has a large angular size while much of the relevant physics takes place on very small spatial scales; the aspect of most interest, the variability of the atmosphere, is due to the presence of strong and intermittent magnetic fields rooted in the photosphere. The **B** field traverses a broad temperature regime from the visible surface to the corona. The wavelengths that need to be observed cover a comparable range, from visible to X-ray, and the spatial scales that need to be resolved range from a fraction of an arcsec at the surface to $> 10^5$ km in the corona; the time scales that need to be studied range from microseconds to years (Golub, 2003). The combination of instruments chosen for the *Hinode* mission has been selected with these considerations in mind.

2. Brief Science Overview

Because the solar outer atmosphere is at a temperature of several million kelvins (MK) its primary emission is in the soft X-ray spectral range. X-ray images display the spatial distribution of this high-temperature plasma, and diagnostics are available to determine the distribution as a function of temperature of the coronal material. The X-ray Telescope (XRT) provides an unprecedented combination of spatial, spectral, and temporal coverage, which will allow a broad range of scientific investigations to be carried out, as listed in Table 1. In addition to the inherent capabilities of the XRT, the spectral data provided by EIS and spectroheliograms and the **B** measurements from SOT/FPP will greatly extend our analytic capability in studies of the structure and dynamics of the solar atmosphere.

The XRT provides several "firsts" in capability in comparison with previous X-ray imagers. Some of these are summarized in the following, and further details are presented in Deluca *et al.* (2005):

Table 1 Scientific objectives of the XRT.

Objective	Representative questions
1. Coronal mass ejections	How are they triggered? What is relation to **B** structure? What is relation between large-scale instabilities and dynamics of fine structure?
2. Coronal heating	How do coronal structures brighten? Are there waves, and do they correlate with brightness? Do loop – loop interactions cause significant heating?
3. Reconnection and jets	Where and how does reconnection occur in the corona? What is its relation to B?
4. Flare energetics	Same as Item 1 above
5. Photosphere – corona coupling	Can a direct connection between coronal and photospheric events be established? How is energy transferred to the corona? What determines coronal transverse fine structure?

– Unprecedented combination of spatial resolution, field of view, and image cadence.
– Broadest temperature coverage of any coronal imager to date.
– High data rate for observing rapid changes in topology and temperature structure.
– Extremely large dynamic range to detect corona from coronal holes to X-flares.
– Flare buffer, onboard storage, and high downlink rate.

2.1. Instrument Flowdown Requirements

The scientific objectives that have been specified for the XRT lead to a set of flowdown requirements, as shown in Table 2. These are the minimum performance requirements needed to meet the objectives; in most cases the as-built XRT exceeds these requirements, as will be discussed in the following.

These requirements lead to the choice of a grazing-incidence (GI) X-ray telescope, with a design chosen to meet the specific requirements. We note that the XRT design provides image quality in the central portion of the field that is as good as manufacturing tolerances allow, so that no further improvement in image quality could be considered. The design and fabrication of the XRT are discussed in detail in Section 3.1.

2.1.1. XRT Temperature Sensitivity

Requirements 3 and 4 in Table 2 may be singled out as involving more of the telescope design elements simultaneously than any of the other requirements. There are a number

Table 2 Instrument flowdown requirements.

Requirement	Definition	Value
1. Exposure time	Shutter-open time	4 ms – 10 s
2. Cadence	Time between exposures	2 s (reduced FOV)
3. T range	Limits of temperature coverage	$6.1 < \log T < 7.5$
4. T resolution	Ability to discriminate temperatures	$\log T = 0.2$
5. X-ray optical resolution	Diameter of 50% enclosed energy	2 arcsec PRF
6. Field of view	Angular coverage of telescope	>30 arcmin
7. WL rejection	Reduction of solar visible light at focal plane	$>10^{11}$
8. Data rate	Maximum bit transfer rate out of XRT	2.4 Mb/s
9. Data volume	Maximum volume per memory fill cycle	60 MB/orbit
10. Coordination X-ray/WL (spatial)	Coalign X-ray and WL	One XRT pixel
11. Coordination XRT/SOT (spatial)	Coalign XRT to SOT	One XRT pixel
12. Coordination XRT/EIS (spatial)	Coalign XRT to EIS	One XRT pixel

of factors involved in determining the temperature range of a GI telescope, and additional factors determine the temperature sensitivity. Design elements include the reflectance of the telescope as a function of incoming photon energy E; the transmission of the entrance aperture prefilters as $f(E)$; the transmission of the focal plane analysis filters as $f(E)$; and the response of the focal plane detector as $f(E)$. We see that each of the major subsystems of the XRT contributes to this requirement. In the following sections we discuss the major subassemblies of the XRT separately. The overall temperature response and temperature discrimination of the XRT will be discussed subsequently.

3. Major Subsystems

The components of the XRT are shown schematically in Figure 1. The XRT is, in principle, a simple instrument: a grazing-incidence optic focuses solar soft X-rays onto a CCD array. Separation between the two is maintained by a stiff, lightweight, low-expansion carbon-fiber-reinforced polymer (CFRP) tube. The front of the telescope is covered with thin filters to reduce the visible light entering the telescope, and additional filters are positioned in the optical path ahead of the CCD camera, along with a shutter assembly. A focus mechanism allows the camera to be moved ± 1 mm along the optical axis. A visible light optic is included in the center of the X-ray mirror to provide aspect information for the XRT and for the *Hinode* mission as a whole. The telescope (*i.e.*, the mirror) is shown schematically in Figure 2. Its design is discussed in detail in Section 3.1.

3.1. Mirror

Grazing-incidence optics used for soft X-ray imaging generally require a minimum of two surfaces to meet the Abbé sine condition that the magnification be constant over the full aperture of the telescope. Wolter used a paraboloid–hyperboloid design, and the more complex Wolter–Schwarzschild design improves image quality slightly by exactly satisfying the condition. Werner (1977) recognized that for a wide-field instrument the field-averaged point spread function (PSF) is a better figure of merit to use, and modern computer polishing methods permit the implementation of designs using high-order polynomial surfaces that deviate from the more standard conic sections. These designs generally trade on-axis image quality for off-axis improvements. This is acceptable because perfect surfaces cannot

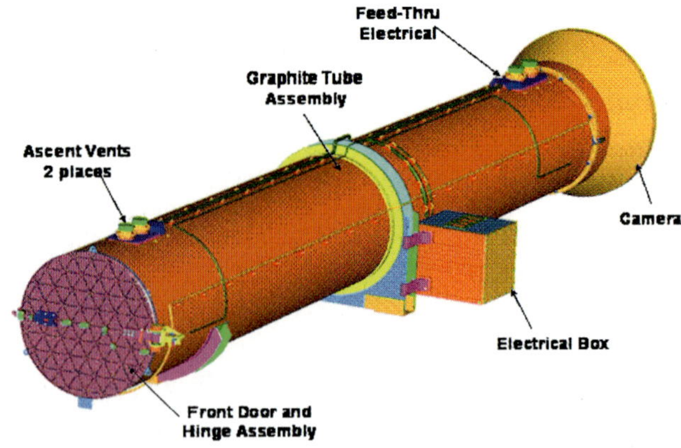

Figure 1 The major components of the XRT; the optics are located inside the entrance aperture door.

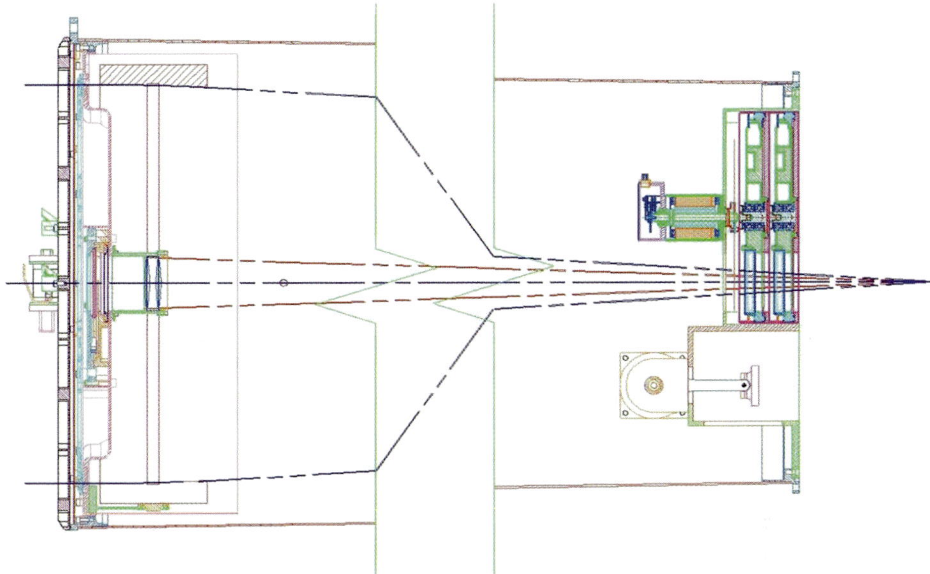

Figure 2 A schematic diagram of the grazing-incidence XRT including the entrance filter assembly and the centrally located visible light telescope. The GI telescope is shown in section (cross-hatched). The shutter, filter wheel, and focus mechanism (US portion) are shown at right. Note that the central part of the CFRP metering tube has been omitted and that the camera is not shown.

be manufactured, so that there is relatively little loss on axis, whereas the off-axis gain can be considerable. In addition, the detector can be positioned slightly out of focus and thereby achieve a better PSF at finite field angle, at the cost of on-axis performance; this was done, *e.g.*, for the *Yohkoh* SXT (Nariai, 1987, 1988; Tsuneta *et al.*, 1991).

The XRT uses a generalized asphere for each mirror element and also has a focus mechanism that allows images to be obtained at the best on-axis (Gaussian) focus and at a range of defocus positions. The variation of image quality versus field angle can be varied by changing the location of the focal plane, as shown in Figure 3.

3.1.1. Goodrich

3.1.1.1. As-built   The telescope as delivered to SAO met all of the design requirements except for knowledge of the focal length, as shown in Table 3. The stated encircled energy diameter requirement was for the optical design itself, and the delivered performance matches the requirement for 2 arcsec imaging.

3.1.1.2. Performance Predictions   Measurements of the as-built telescope at Goodrich were used to calculate the expected optical performance of the XRT. Figure 3 shows the predicted as-built RMS point response function (PRF) versus field angle at different field angles. Comparison with the measured PRF will be presented in Section 4 of this paper.

The size and shape of the PRF are functions of position in the field and of focal plane location along the optical axis. This is illustrated by Figure 4 (top), which shows the PRF at several off-axis locations for a focal plane located 200 μm ahead of the best Gaussian focus. The corresponding encircled energy plots are shown in Figure 4 (bottom). These plots are again predictions based on the measurements of the as-built XRT.

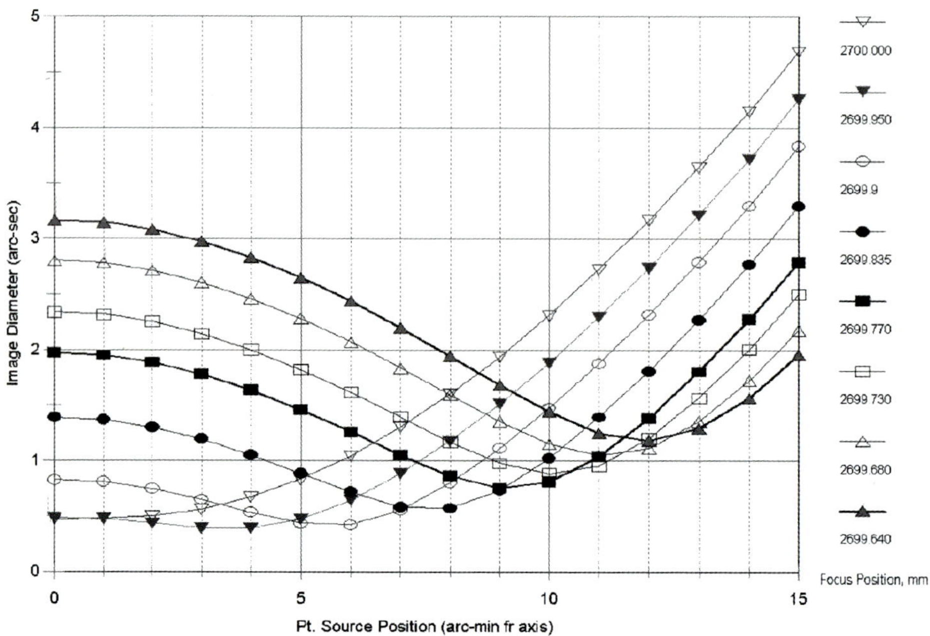

Figure 3 The predicted point response function of the XRT as a function of focal plane location, based on Goodrich measurements of the mirror surface shape.

Table 3 As-built XRT performance parameters.

Parameter	Requirement	As-built
Optical design	Single mirror pair	Generalized asphere
Wavelength range	6 – 60 Å	Bare zerodur
Entrance diameter	341.7 ± 0.1 mm	341.7 mm
Focal length	2708 ± 2 mm	2707.5 mm
Focus knowledge	±0.050 mm	±1.4 mm
Field of view	35 arcmin	Optimized over 15 arcmin
Encircled energy	68% at 0.5 keV	68% at 0.56 keV
(diameter)	1.57 arcsec	2.3 arcsec
Effective area	1.0 cm^2	1.9 cm^2

The focal length of the XRT (and of GI telescopes in general) is dependent on field angle because of focal plane curvature. The predicted values at field angles of 0, 5, 10, and 15 arminutes are 2707.5, 2707.0, 2706.8, and 2707.2 mm, respectively. The spot centroid for these off-axis point source images is located ≈0.78553 mm/arcmin from the on-axis field center; this quantity is the plate scale.

The measured performance of the XRT will be discussed in Section 4 where we present results of the extensive calibration and testing that were carried out at the X-Ray Calibration Facility (XRCF) of the Marshall Space Flight Center.

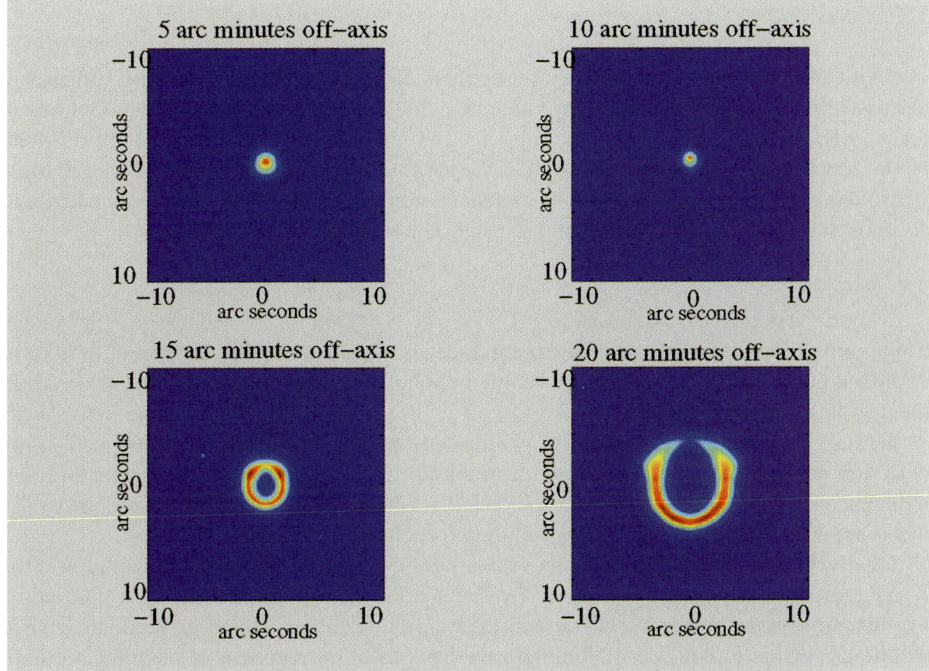

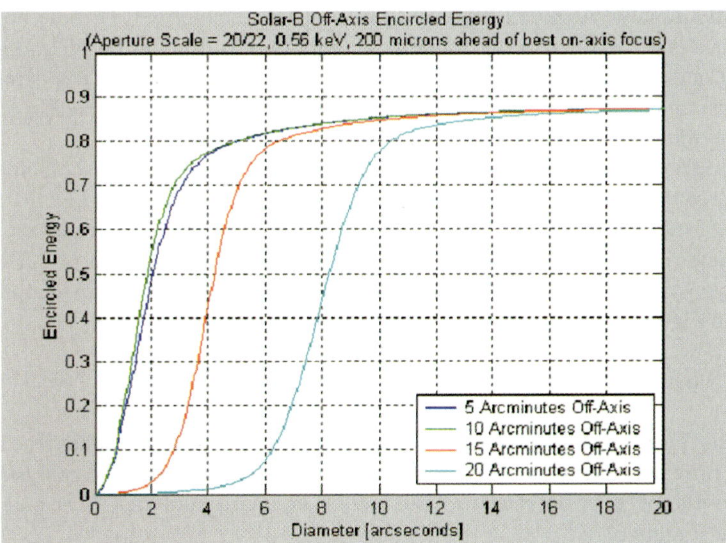

Figure 4 The predicted PRF of the XRT for off-axis angles at a focus position 200 µm forward of best focus. The top image shows the focal spot and bottom figure shows the corresponding encircled energy curves (Courtesy Goodrich – Danbury).

3.2. Filters

The XRT utilizes both entrance aperture prefilters and focal plane analysis filters; these are discussed in the next two sections. We note here that the visible-light-blocking requirement for the XRT is 10^{-12}, which translates into a requirement for the prefilter and analysis filters of 10^{-6} each. This requirement determines the minimum acceptable thickness for each filter. It also drives the mechanical design when launch vibration and acoustic loads are taken into account, since the filters must remain light-tight into orbit.

3.3. Prefilters

Thin prefilters cover the narrow annular entrance aperture of the XRT. These prefilters serve two main purposes: (1) to reduce the visible light entering the instrument and (2) to reduce the heat load in the instrument.

Taking into account the amount of visible light attenuation needed to allow soft X-ray detection at the focal plane CCD, and also taking into account the available methods for reducing visible light entering the telescope, we find that Requirement 2 is automatically met when Requirement 1 is met. We will therefore discuss only the first item.

Ideally, the prefilter will reduce the visible light as much as possible and will reduce the X-ray throughput as little as possible. Experience has shown that the material best suited for this purpose is aluminum and we will treat only this choice. However, there are several additional factors to consider: Aluminum oxidizes from the moment of manufacture until launch. The rate is strongly affected by humidity, so that handling in dry nitrogen is important. Additionally, Al filters exposed to full Sun in vacuum will become hot. Thermal conduction along the filter to the frame that supports it will cool the filter, and the path length is a major factor in determining the temperature on orbit. Finally, to strengthen the filter against launch loads, the aluminum can be mounted on a mesh, on a thin organic film such as polyimide, or on both.

The severity of the vibration and acoustic loads for the *Hinode* launch, combined with the availability of thin, strong, highly transmissive polyimide from the Luxel Corp. determined the choice of substrate thickness. The amount of Al need to provide the required visible light blocking determined the Al thickness. The XRT entrance filters consist of 1200 Å Al and 2500 Å of polyimide, with an estimated 100 Å of Al_2O_3. A photo of one entrance filter is provided in Figure 5.

3.4. Focal Plane Analysis Filters

The analysis filters serve two purposes: (1) to reduce the visible light reaching the focal plane and (2) to provide varying X-ray passbands for plasma diagnostics. The analysis filters are held in two filter wheels, operated in series and located in front of the CCD camera, near the focal plane (Figure 6).

The analysis filters, differing in thickness by nearly a factor of 10^4, greatly extend the dynamic range of the XRT, For faint targets, which also tend to have relatively low temperatures, the thinner filters are used. For flare observations the thickest filters are used. In addition, there is a focal plane filter used with the visible light telescope, which is operated separately from the X-ray channel. This glass filter may also be inserted during X-ray observations in the rare event that the flare intensity exceeds the instrument saturation level.

Inserting the analysis filters into the optical path alters the throughput of the telescope as a function of wavelength. Combined with software that calculates the spectral emissivity of

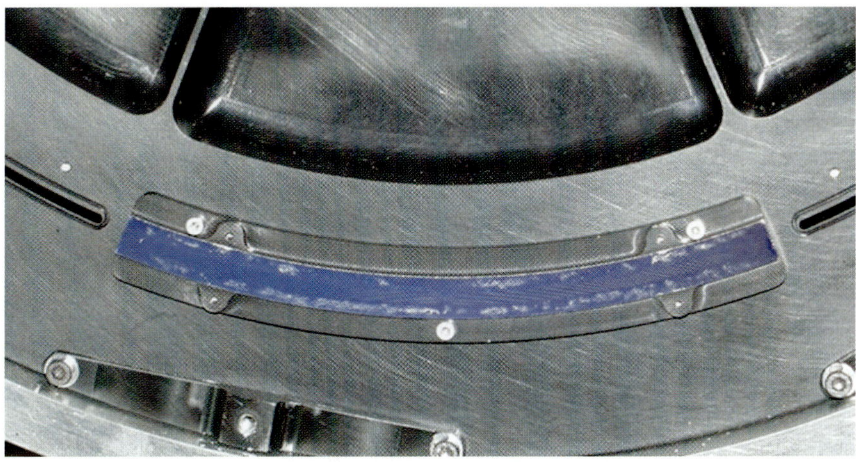

Figure 5   One of the six annular aluminized polyimide entrance filters of the XRT.

Figure 6   One of the two XRT filter wheels, each of which holds five filters and has one open position.

the coronal plasma as a function of its temperature (Smith *et al.*, 2001), we may calculate the response of the XRT for the various analysis filters. This calculation is shown in Figure 7, in which the throughput of the telescope is shown for an assumed source with constant emission measure (amount of material) as a function of temperature of 1×10^{30} cm^{-5} over the entire XRT sensitivity range. The units are erg cm^{-2} s^{-1} at the focal plane, so that the CCD response is not included. These curves represent the basic first step in the quantitative analy-

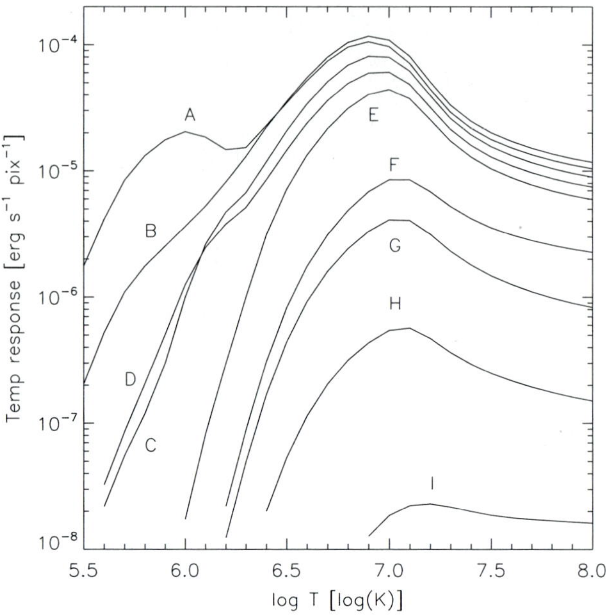

Figure 7 The total XRT temperature response, for all of the X-ray focal-plane filters. Each curve plots the combination of the total instrument response (as a function of wavelength) with a coronal plasma emission model (ATOMDB/APEC) for a columnar emission measure of 10^{30} cm^{-5}. The labels indicate which filter is in the path, as follows: A = Al-mesh, B = Al-poly, C = C-poly, D = Ti-poly, E = Be-thin, F = Be-med, G = Al-med, H = Al-thick, and I = Be-thick.

Table 4 Physical characteristics of the XRT focal plane analysis filters.

Filter ID	Material	Thickness (Å)	Filter support	Thickness (Å)	Oxide	Thickness (total, Å)
Al-mesh	Al	1600	–	82%	Al_2O_3	150
Al-poly	Al	1250	Polyimide	2500	Al_2O_3	100
C-poly	C	6000	Polyimide	2500	N/A	N/A
Ti-poly	Ti	3000	Polyimide	2300	TiO_2	100
Be-thin	Be	9E4	N/A	N/A	BeO	150
Al-med	Al	1.25E5	N/A	N/A	Al_2O_3	150
Be-med	Be	3.0E5	N/A	N/A	BeO	150
Al-thick	Al	2.5E5	N/A	N/A	Al_2O_3	150
Be-thick	Be	3.0E6	N/A	N/A	BeO	150

sis of XRT images; a typical next step might be the type of analysis described in Section 5.4 to determine the distribution of the EM in a target region of interest.

3.4.1. Filter Specification

A summary of the nominal physical properties of the focal plane analysis filters is provided in Table 4. The measured values for the component thicknesses are discussed in the next section.

3.4.2. Palermo Testing

Several of the flight focal plane filters for the XRT were tested at the X-Ray Astronomy Calibration and Testing (XACT) facility of INAF – Osservatorio Astronomica di Palermo.

Figure 8   The XRT shutter is a thin rotating blade with two narrow slits for 1 and 8 ms exposures, plus a slot for exposure times ≥ 44 ms.

The goal of these calibrations was to determine the spatial uniformity of the filters and the transmission properties (Barbera *et al.*, 2004). Of the nine filters tested, seven are installed in the XRT, since several were damaged in shipping and needed to be replaced. The results show that the spatial uniformity is 2% or better for the metal on polyimide filters and better than 3.3% for the single metal filters. The transmission tests showed that the results were within $5-10\%$ of the predicted values.

After launch of the XRT the witness samples for the flight filters (entrance and focal plane) will be measured for transmission at the XACT. This will provide the final transmission calibrations.

3.4.3. XRCF Testing

The transmissions of the focal plane filters listed in Table 4 were also measured during the end-to-end test at the XRCF. The measured transmissions are within $5-20\%$ of the predicted values for all of the exposures with a high photon flux. Table 6 shows the sources that we had available to us for these tests, and Table 7 shows the comparison of measured and predicted transmissions.

3.5. Shutter

The focal plane shutter used in the XRT is a modified version of the TRACE shutter (Figure 8). It can be operated in two modes: a continuous sweep at fixed rotation speed or a start–stop mode in which the large opening is brought into the path and kept there for the desired length of time. The narrow openings may be employed in a multiple-pass mode, so that a large set of exposure times can be used. The set of exposure times chosen for initial operations is given in Table 5.

Table 5 Initial table of available XRT exposure values.

ID#	τ_{exp} (s)	ID#	τ_{exp} (s)
0	0.001	18	0.71
1	0.002	19	1.00
2	0.003	20	1.41
3	0.004	21	2.00
4	0.005	22	2.83
5	0.008	23	4.00
6	0.012	24	5.66
7	0.016	25	8.00
8	0.024	26	11.3
9	0.032	27	16.0
10	0.044	28	22.6
11	0.063	29	32.0
12	0.086	30	45.2
13	0.125	31	64.0
14	0.177	32	64.0
15	0.250	33	64.0
16	0.354	34	64.0
17	0.500	35	64.0

3.6. WL Telescope

The XRT includes a visible light imager coaxial and confocal with the X-ray telescope. It is a simple achromat designed to image the Sun in the 400–500 nm band with 2 arcsec spatial resolution. The lens materials, fused silica and SF16, are chosen for their insensitivity to radiation darkening and are used to correct axial color between 405 and 495 nm; the design also corrects for spherical aberration at 430 nm.

The nominal focal length of the doublet is 2705 mm, and the aperture is 50 mm, giving an f-number of f/54. The optic is mounted in a manner that permits adjustment along the optical axis so that the visible light focal plane is positioned coincident with the as-built X-ray focal plane, to within the depth of focus of the XRT. The WL depth of focus is substantially larger than the XRTs.

The WL Telescope is fitted with an entrance aperture filter manufactured by the Andover Corporation and having a passband centered on the G band at 430.7 nm with a FWHM of 18.9 nm. A similar filter is placed in one of the focal plane filter wheels, with a tested central wavelength of 430.3 nm and FWHM of 17.7 nm. The peak transmissions of the filters are 49.3% and 29.6%, respectively, including a built-in neutral density coating on the focal plane filter for off-band rejection. The additional neutral density filter, of ND $=$ 1.3, was selected to provide a nominal exposure time of 1/100 s. The correct ND value was chosen by calibrating the WL Telescope at Williams College by using their 0.6 m solar telescope and obtaining images at varying elevations, so that an extrapolation to zero air mass could be obtained.

3.7. Visible Light Imager and X-Ray Telescope Confocality

The XRT and visible light imager (VLI) share a CCD camera and a focus mechanism that has a range of motion of ± 1 mm. Therefore, to achieve the required optical performance, it is necessary that their respective best on-axis focus positions both lie within the range of the focus mechanism. To reasonably minimize cadences and the frequency of mechanism motions, it is desirable that the best on-axis focus position for the VLI lie within 250 μm of that for the XRT. This value is substantially smaller than the depth of focus of the VLI, so that refocusing in switching from XRT to VLI would not be needed.

Observations were taken of a visible light penray lamp at the same distance along the facility axis as the X-ray source. Measurements were taken across the range of focus mechanism positions, with the VLI shutter in both the open and closed states. At this point in the end-to-end test, the telescope configuration included the finite source spacer, but not the entrance filters. Therefore, images taken with the VLI shutter open registered light traversing both the XRT and VLI optic. Images taken with the VLI shutter closed registered light traversing the XRT optic. To isolate an image focused solely by the VLI optic, each closed image is subtracted from its corresponding open image. This subtraction also compensates for dark current in the image.

The images of the visible light source were fitted with a 2D Gaussian model, which is valid for focus positions near best focus. The RSS Gaussian widths are then plotted as a function of focus position (Figure 9). The best on-axis focus position is defined to be where the RSS Gaussian width is a minimum. However, analysis indicates that, over the range of motions available in the XRT, the data are well fit with a horizontal straight line. This result corresponds to an instrument that has a broad depth of focus, in which case all (on-axis) focus positions produce images with equivalent focus.

The VLI data are consistent with the prediction that the VLI would have a depth of focus (2.2 mm) broader than the range of motion of the focus mechanism (1.0 mm). (Note that the size of the spot in Figure 9 includes the finite size of the lamp used to produce the WL

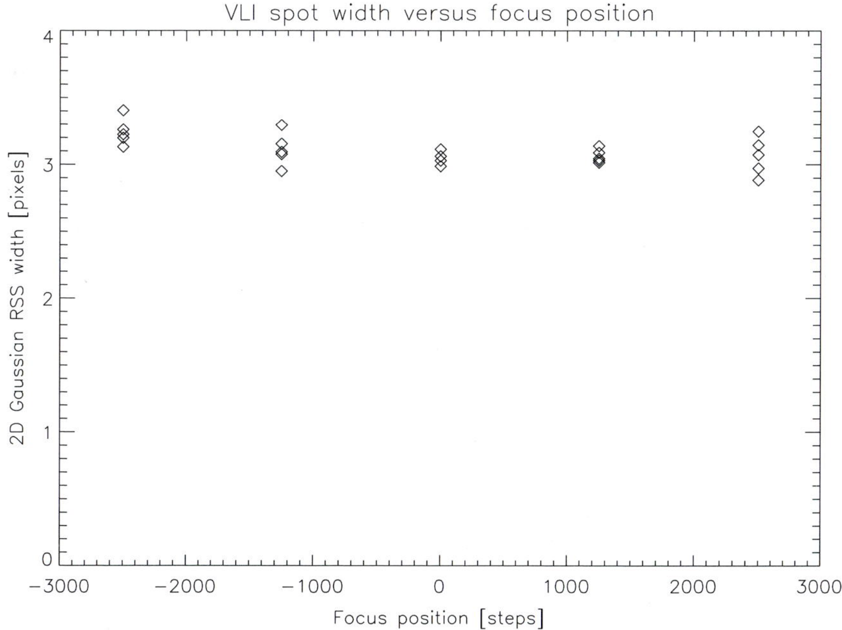

Figure 9 The spot size of a visible light source imaged through the visible light imager (VLI) as a function of distance along the optical axis of the XRT.

image at the XRCF.) Therefore, any VLI focus position within 250 μm ($\approx$500 steps in the diagram) of the XRT best-focus position will give acceptable focus of the VLI.

3.8. Visible Light Imager and X-Ray Telescope Coalignment

The visible light source was placed adjacent to the X-ray source at the same horizontal distance (530.6 m) from the front of the XRT instrument along the facility axis, but laterally offset by 14.1 cm.

For the X-ray measurements, the Cu $-$ L source was used. Normal images were taken at the best-focus position, and the XRO peak flux centroids were located to ±0.5 arcsec, *i.e.*, one pixel on the CCD. For the VLI measurements, normal images were taken at the nominal 430.5 nm focus position, and the VL peak flux centroids were also located to ±0.5 arcsec.

The measured offset between the XRT and VLI centroids was corrected for the physical offset of the sources and an estimate of the error in the measure of the offest. The VLI and XRT axes net alignment offset is 17.0 ± 5.0 arcsec, with the uncertainty dominated by the $\approx1/4''$ uncertainty in the location of the visible light source. This alignment accuracy is well within the required coalignment tolerance of one arcmin. The tighter requirement on *knowledge* of the coalignment is determined after launch during the commissioning phase of the satellite.

4. XRT Mirror Imaging Performance

Two major test sequences of the XRT were carried out at NASAs XRCF. In the first (mirror calibration test), the imaging properties such as point response function, effective area as a function of wavelength, and off-axis response of the XRT were determined. The XRT was then returned to the XRCF after assembly into flight configuration, specifically to establish the focal length via an end-to-end test. Image performance data from both the mirror calibration and end-to-end test are presented in this section. The total telescope throughput is discussed in the following section. Details of the CCD camera calibration are discussed in the companion XRT camera paper (Kano *et al.*, 2007).

4.1. Test Plan

The XRCF consists of an X-ray source located at one end of a 518 m vacuum pipe, with a large vacuum chamber at the other end of the pipe to hold the test article and associated test equipment. Our tests used an electron impact point source, in which an electron beam is focused onto a target, at a voltage chosen to excite a characteristic line; typically K-α or L-α lines are chosen, for strength and spectral purity. A thin metallic filter, usually of the same material as the target, is then placed in the path to filter out unwanted X-ray energies while permitting the characteristic X-rays to pass. The primary lines used in this test are listed in Table 6.

The mirror calibration tests carried out were as follows:

Focus Determination: The mirror is aligned with the source $-$ detector line. A scan through the focus range was performed and 2D Gaussian curve fits were applied to images. The best on-axis focus (BF) was located by the minimum in the polynomial fit of the RSS sigma width values for the FWHM. The focus position that provides the best average imaging across the field of view (FOV) is called the FOV Optimized Focus (OF), defined as the position where the on-axis FWHM is equal to $2''$. This yields the largest extent of field angles meeting the resolution requirement.

Table 6 Available X-ray lines for the XRT calibration at the XRCF.

Line	Energy (keV)	Wavelength (Å)
C – K	0.277	44.7
O – K	0.525	23.6
Cu – L	0.933	13.3
Al – K	1.49	8.3
Mo – L	2.29	5.4

On-Axis PSF and Encircled Energy: Characterization of the performance of the XRT mirror was performed after completion of the optical axis alignment of the mirror to the X-ray beam. Data were collected using the Cu – L source from both the CCD and proportional counters. The combination of these data was used to find the PSF and encircled energy.

On-Axis Effective Area: A flow proportional counter (FPC) with a wide pinhole aperture was used to alternately measure the photon count rate through the effective area of the X-ray optic and through the unobstructed visible light aperture along the same optical axis. The ratio of the on-axis effective area to the calibrated pinhole aperture area is equal to the ratio of the photon count rate through the respective areas. Independent measurements of the source flux rate provided normalization against time variations. Proportional counter data were taken with the five different energy sources (Table 6).

Off-Axis PSF and Encircled Energy: This test was performed with the Cu – L (0.933 keV) source, using an off-axis configuration of 15.6 arcmin. The X-ray beam center was measured with an FPC and an array of calibrated pinhole apertures from 20 μm to 20 mm. These data were used to characterize both the off-axis PSF and encircled energy performance of the XRT optic.

Off-Axis Effective Area: The XRT off-axis effective area was measured at an angular displacement of 15.6 arcmin (immediately subsequent to the off-axis PSF measurements). The 20 mm pinhole aperture and an FPC detector were used to sample the beam center for both the Cu – L (0.933 keV) and O – K (0.525 keV) X-ray sources, in conjunction with measurements through the VLI. In other respects, the test procedure was identical to that described for the on-axis effective area measurements.

Wings of the PSF: To determine the contribution of the wings of the PSF, 89 normal and dark images were collected at best on-axis focus. A subset of these images were calibrated and summed. Fluxes within annuli approximately 3 CCD pixels thick were summed with the midpoint radius of each annulus expanding toward the edge of the CCD in discrete steps of 3 CCD pixels.

Thermal Response of the Telescope: Images were collected at three different chamber temperature configurations: at 15°C, at 22°C, and with a 1°C temperature gradient around the mirror circumference. Analysis of any observed variation in the PSF FWHM was made.

4.2. Best On-Axis Focus PSF

The BF was found as described in the previous section. During the mirror calibration tests data involving subpixel motions were collected to help determine the azimuthal location of best focus. In this procedure, images were taken after the CCD position was shifted 7 μm ($\approx 1/2$ pixel) in both the $\hat{z}$ and the $\hat{y}$ directions. An estimate of the PSF was derived from these data via an interpolation method (Figure 10), after which corrections for the finite

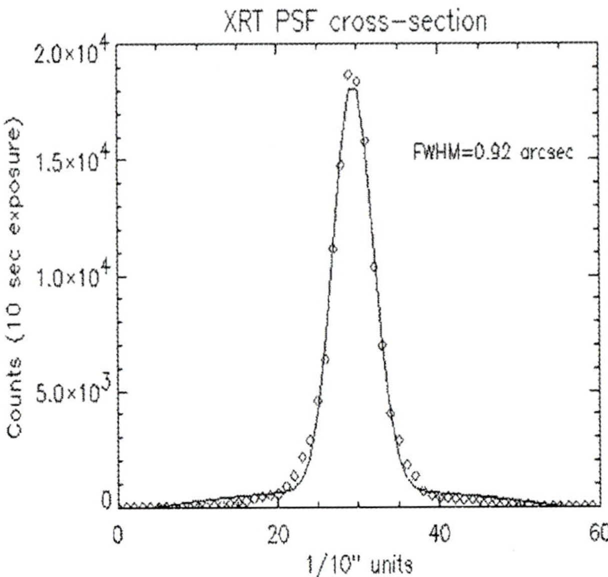

Figure 10 The PRF of the XRT before correction for the finite source distance during XRCF testing and for the deformations from gravity. The FWHM is less than one XRT CCD pixel.

source distance and the effect of gravity were applied to derive the PSF as it will appear in flight configuration (Figure 11).

After the PSF is constructed, annuli of successively increasing inner radii were formed, with the peak of the PSF taken as the origin of all of the annuli. The fluxes contained within these 2-pixel-thick annuli were summed and plotted as a function of diameter of the annuli. The resulting plot is shown as one of the curves in Figure 12. This plot shows that the mirror meets the NASA requirement that 50% of the encircled energy be contained within a diameter of 27 μm (2 arcsec).

The PSF was also measured via a series of pinholes positioned on axis at the plane of best focus, with a proportional counter located behind the pinholes. These measurements confirmed the size and shape of the inner core of the PRF and were also used to measure the large-angle wings of the PRF; those results are presented in Section 4.3.

The PSF as measured contains two contributions that are not present on orbit: The XRCF source is of necessity located at a finite distance from the telescope, and the optic is distorted from the effect of gravity. We have modeled both of these effects and corrected the measured PSF, with the result shown in Figure 11.

4.2.1. Mirror Contribution

The experimentally determined encircled energy function of the XRT is shown in Figure 12, which compares the prediction with two measurements, one using a CCD and one using a nested sequence of pinholes. The only discrepant data point is that using the large (50 μm) pinhole; this is being investigated. The CCD contribution to the instrument PRF is discussed in the companion paper in this volume.

4.2.2. Performance across the FOV

Images taken at Cu $-$ L (0.933 keV) across the field of view of the XRT are used to determine the PSF when the camera is positioned at the plane of best focus. The results are summarized

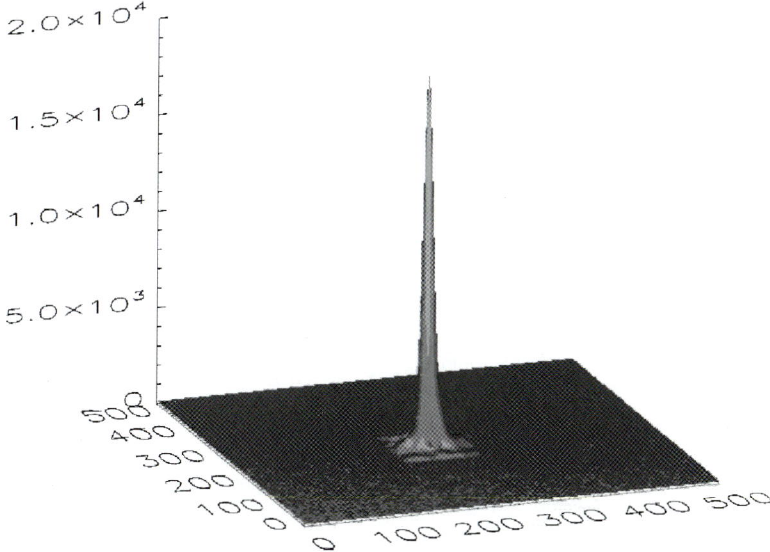

Figure 11 The PRF of the XRT corrected for the finite source distance during XRCF testing and for the deformations from gravity. The FWHM is ≈0.8 arcsec, less than one XRT CCD pixel.

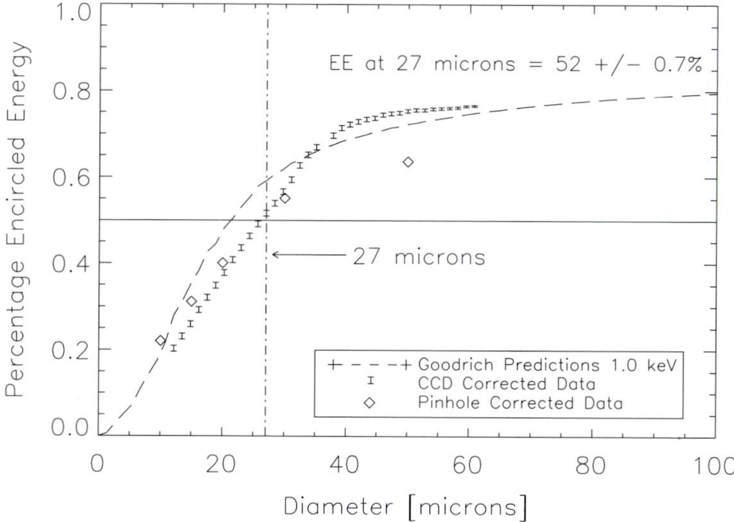

Figure 12 The predicted vs. measured encircled energy function of the XRT for the on-axis Gaussian focus position. The dashed line is the Goodrich prediction and the diamonds and brackets indicate the experimental values.

and compared to a curve showing the variation in RMS spot size versus field angle obtained using a ray trace program that takes the measured telescope surface figure. Comparison of the spot size data obtained at the XRCF after finding the location of best focus to the predictions is shown (Figure 13). We find excellent agreement between the predicted performance and the measured values.

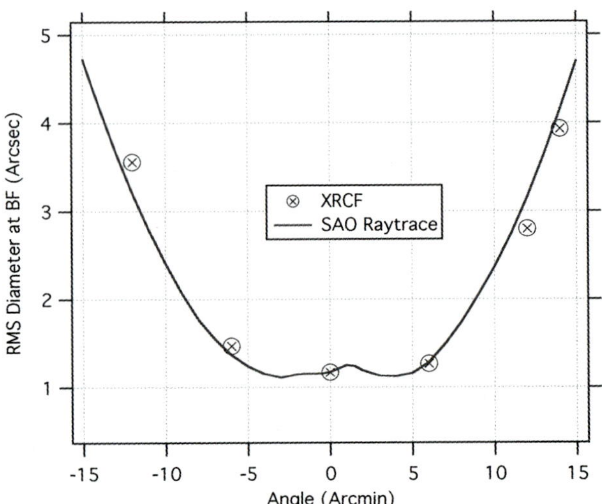

Figure 13 The measured vs. predicted off-axis spot size in test configuration (*i.e.*, without the 1G and finite source distance corrections).

4.3. Wings and Scattering

The wing response is defined as the ratio of the off-axis flux to the on-axis flux for a point source. The off-axis angle must also be specified and for the XRT the requirement was specified for an annular ring of width 1 arcsec located at a radius of 60 arcsec. The wing response was measured at the XRCF for the full range of angles from 0 to slightly more than 60 arcsec by using the CCD images for the bright inner portion of the PSF and the pinhole measurements for the faint outer portion. Ten long-exposure CCD images were averaged for the core determination and the result was spliced onto the pinhole data in their overlap region. The wing response of the XRT PSF (Figure 14) was sampled with an FPC by using 100 and 300 μm calibrated pinhole apertures, at a range of positions between 0 and 1000 μm from the core center. A 2D Lorentzian model was fit to this dataset. The PSF core was sampled with an FPC by using a 10 μm pinhole, in a 7 × 7 array of positions within 22 μm of the core center. A 2D Gaussian model was fit to this latter dataset. The Gaussian core and Lorentzian wing models were matched at 13 μm, and together they were normalized to the core peak value. The measurements shows that the scattering is $<10^{-5}$ at 1 arcmin off-axis, at an energy of 0.93 keV.

5. Throughput

5.1. Mirror Effective Area

The effective area of the XRT depends upon the geometric area and the reflectance of the mirror surfaces. The latter quantity is wavelength dependent and enters twice because the XRT is a two-bounce telescope. The predicted effective area agrees well with the measured area, as shown in Figure 15. The effective area at an off-axis angle of 15.6 arcmin was also measured, as shown in the figure; the lower reflectance indicates the degree of vignetting at this angle. The overall XRT response also includes the entrance filter and the focal plane analysis filter transmissions and the CCD efficiency.

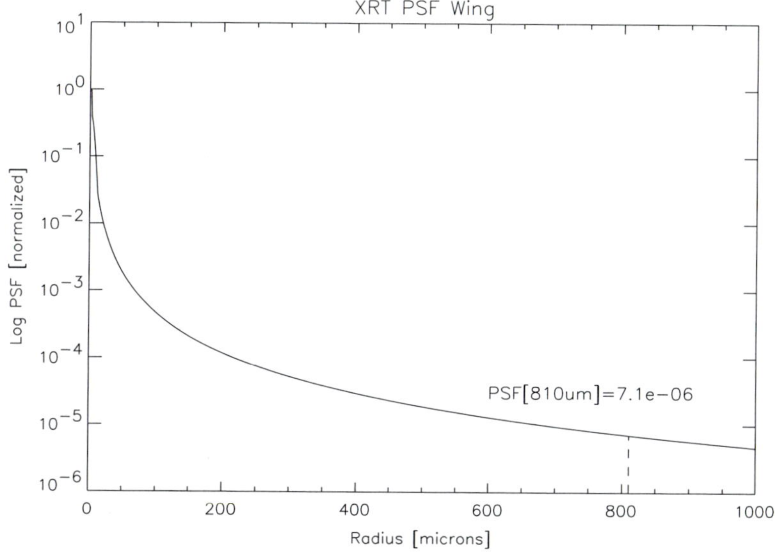

Figure 14 The wing response of the XRT PSF at a range of positions between 0 and 1000 μm from the core center. A 2D Lorentzian model was fit to this dataset; see text for details.

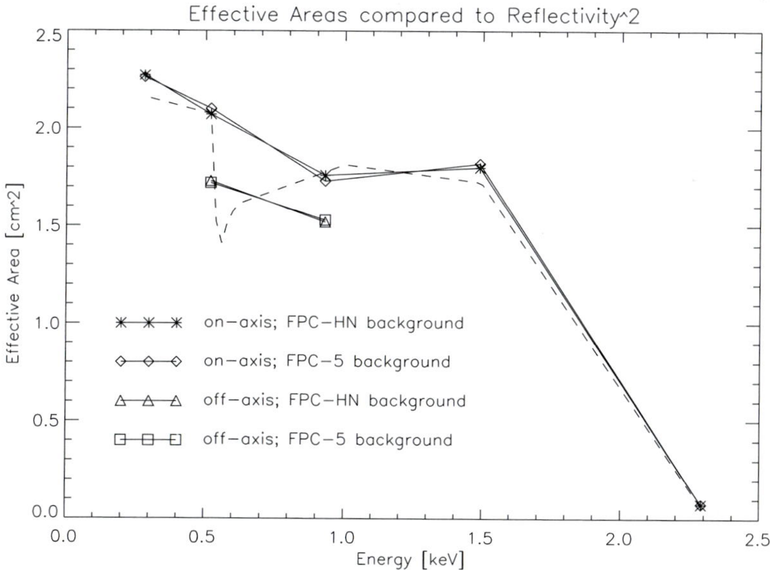

Figure 15 The effective area of the XRT optic as a function of energy, both on and off axis. The predicted on-axis areas for two reflections on the mirror (dashed line) are consistent with the on-axis measurements (asterisks and diamonds on solid lines). Off-axis measurements are also shown (triangles and squares on solid lines) for a field angle of 15.6 arcmin.

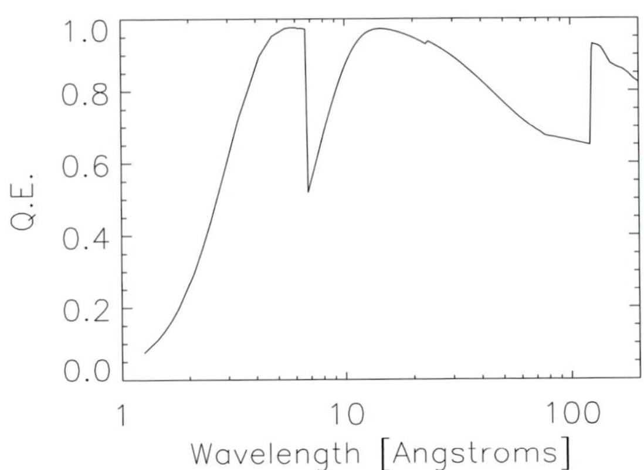

Figure 16 The nominal quantum efficiency curve for the XRT CCD camera, shown here for illustrative purposes only.

Table 7 Analysis filter measured (upper) vs. predicted (lower) transmissivities.

Emission line	C – K	O – K	Cu – L	Al – K	Mo – L
Energy (keV)	0.227	0.525	0.933	1.49	2.5
Filter	Measured transmission ± uncertainty				
	Predicted transmission				
Al-mesh	23.8 ± 6.9	80.8 ± 8.5	93.7 ± 6.2	94.5 ± 5.2	77.4 ± 7.0
	21.2	72.1	91.9	97.6	92.0
Al-poly	24.0 ± 8.3	51.5 ± 8.6	77.4 ± 7.0	94.5 ± 5.2	89.6 ± 7.3
	19.4	50.3	82.0	94.6	91.2
C-poly	64.1 ± 5.5	7.9 ± 4.8	60.6 ± 4.8	79.2 ± 3.0	94.1 ± 7.0
	65.3	10.0	57.9	86.3	95.6
Ti-poly	41.8 ± 7.2	5.5 ± 3.2	33.9 ± 3.1	68.8 ± 4.4	91.0 ± 6.4
	34.0	3.1	33.0	71.7	89.3
Be-thin	5.6	–	26.1 ± 3.5	77.8 ± 5.6	90.7 ± 6.6
	0.0	0.1	27.8	72.6	91.1
Be-med	6.8	–	4.5 ± 2.7	48.9 ± 6.8	77.2 ± 5.3
	0.0	0.0	1.5	32.6	73.7
Al-med	–	–	2.3 ± 1.7	22.6 ± 4.4	2.5 ± 1.9
	0.0	0.0	0.6	23.6	0.2
Al-thick	–	–	–	5.1 ± 2.1	–
	0.0	0.0	0.0	6.0	0.0
Be-thick	–	–	–	–	7.8 ± 2.9
	0.0	0.0	0.0	0.0	6.7

5.2. Filter Transmission

As a check on the thickness of the analysis filters, their transmission at X-ray wavelengths was tested at the XACT facility in Palermo by using the same set of emission lines available for the XRCF testing; the energies associated with these emission lines were shown in Table 6.

The results of the measurements are shown in Table 7. The uncertainties in the measured transmissions are dominated by counting statistics. For some of the thick filters measured with the weaker emission lines, the time available for testing determined the number of photons that could be collected, so the error bars are relatively large. We see that, to within the measurement uncertainties, there is acceptable agreement in all cases between the predicted and the measured values.

5.2.1. CCD Quantum Efficiency

The calibration of the XRT camera is described in detail in a companion paper in this volume. For reference purposes we provide a calibration curve here, Figure 16, since this is part of the throughput calculation for the XRT instrument as a whole.

5.3. Total Telescope Throughput

The total throughput is the convolution of all relevant terms, including prefilter transmission, mirror effective area, analysis filter transmission, and CCD quantum efficiency. These contributions are summarized as an effective area, and the curves for all of the nine XRT filter channels are shown in Figure 17.

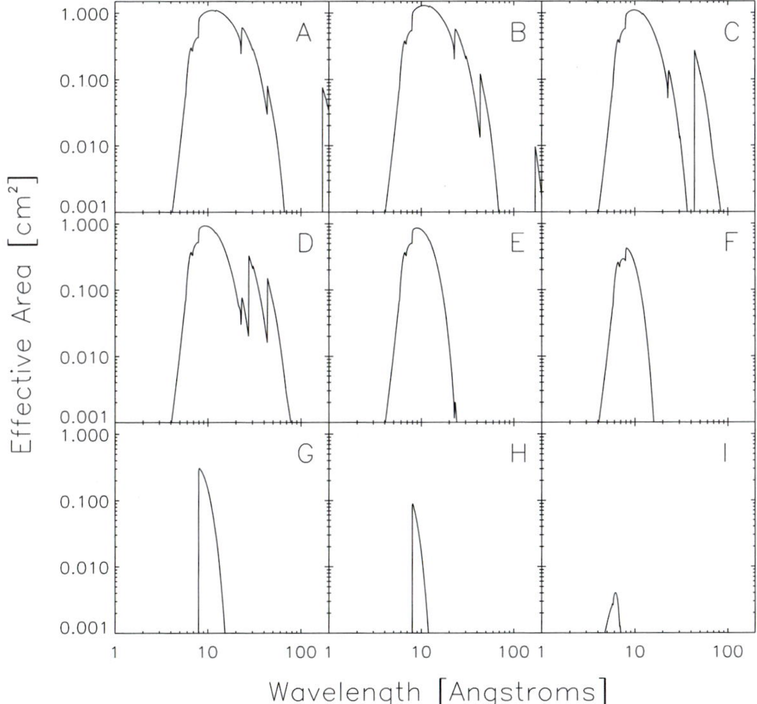

Figure 17 The total telescope throughput of the XRT for each of the nine X-ray filter channels. The labels indicate the filter channels as follows: A = Al-mesh, B = Al-poly, C = C-poly, D = Ti-poly, E = Be-thin, F = Be-med, G = Al-med, H = Al-thick, and I = Be-thick.

5.4. DEM Analysis

The next step after one is able to see the coronal structures is to determine their physical basic properties. Among these is the measurement of the amount of material present in the corona as a function of temperature, the differential emission measure, or DEM. The procedure we use for finding the best-fitting DEM for a given set of observations in several spectral channels has been described in Weber *et al.* (2005). We consider a set of images taken of an active region (AR) and we estimate the DEM in a given pixel. Our procedure then produces an iterative least-squares fit to the observations using a DEM represented by a spline with evenly spaced knots in $\log(T_e)$ space. With the forward modeling approach, we assume a differential emission measure and compare the predicted observations for each filter with the real observations, iterating the DEM until an acceptable fit is found.

The corona is known to be highly inhomogeneous in temperature, density, and magnetic field – the isothermal approximation is often inadequate for describing the optically thin solar atmosphere across length scales comparable to the span of an XRT pixel. The actual DEM distribution in an active region is thus expected to include material across a wide temperature range. We analyzed our DEM procedure using a realistic DEM model that is included in the CHIANTI database to evaluate the methods employed and to establish the number of observing channels needed to reproduce the input data. There is, in principle, no limit to the complexity of the model DEM that could be chosen. However, the physics of the situation – primarily the Boltzmann width of the spectral lines, causing them to be formed over a fairly wide temperature range in the corona – provides a fundamental limitation to the resolving power of any spectroscopic analysis (Craig and Brown, 1976). The amount of structure present in this model is reasonable for these tests.

Figure 18 suggests how well the input DEM can be reconstructed as a function of the number of observing channels used. In the left plot, four XRT channels have been used to perform fits. The figure shows the model AR DEM (solid line with two humps), the distribution of fitted DEMs (gray scale), and the median values of the 100 DEM runs (diamonds).

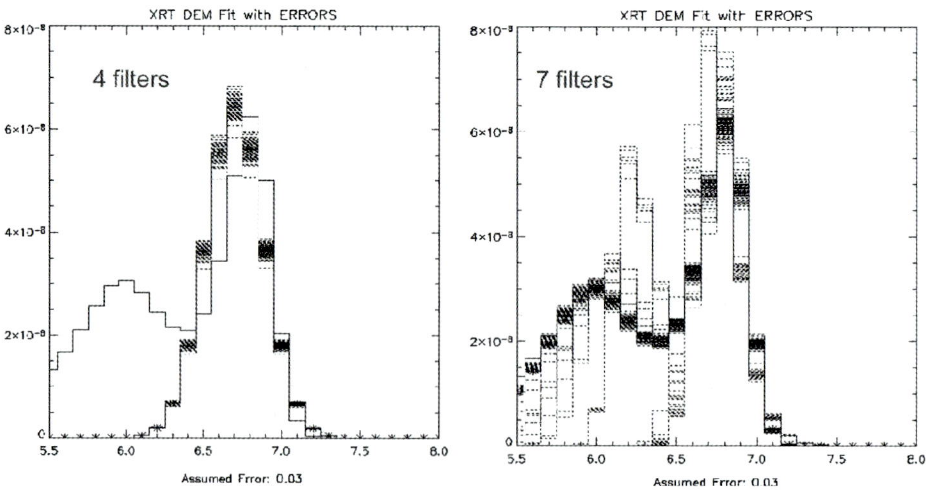

Figure 18 Two attempts at reconstructing an input model (solid histogram) of the DEM from a solar active region: (left) using four XRT channels produces a poor fit; (right) using seven channels yields an acceptable fit. Each panel shows 100 runs, with noise added for realism.

The DEM is fitted over the log temperature range $5.5 - 8.0$ and 3% noise is assumed. These relatively high-T XRT filter channels determine the presence of the hotter peak of material, as indicated by the convergence of the median fit to the model DEM curve, but fail to detect the cooler material. The narrow uncertainty bands indicate that the fits are robust or, in the words of one author, "reliably bad."

In the right plot of Figure 18, the same model is fitted with seven channels; that is, we have included thinner filters in the analysis. It is obvious that the fitted DEMs more accurately reproduce the model DEM curve across the entire temperature range. Even though the uncertainty bands are not as constrained as in the left plot of Figure 18, they adequately indicate the presence and temperature of the cool component. To achieve good results in DEM reconstruction with XRT data, it is thus important to have observations in many (independent) channels.

5.4.0.1. The Number of Channels The physics of ionization fraction formation under coronal conditions combined with the range of temperatures found in the corona leads to a definite requirement on the number of independent channels that need to be recorded to reconstruct the emission measure distribution. The narrowness of the temperature range over which a typical ionization state is formed in the corona, combined with the requirement to have complete but nonredundant coverage, means that at least six channels must be used. This is shown explicitly in Figure 18, where we examine how the removal of even a single channel affects the quality of the DEM reconstruction for a typical active region.

DEM reconstruction thus requires observations in at least six independent channels, most of which should record a useful number of counts. This is the major reason for the choice of up to nine channels, spanning a large temperature range, in the XRT.

This analysis indicates that DEM determinations are data intensive, since they require that a large number of channels be used. Because telemetry is limited, as is onboard storage, and because some scientific objectives require high cadence (*i.e.*, the use of a single channel or a small number of channels), the multichannel DEM programs will need to be specifically scheduled when needed.

6. Conclusions

The X-ray telescope for *Hinode* is the highest resolution solar X-ray telescope ever flown (TRACE is EUV). The optical design and mirror quality ensure excellent imaging performance across the FOV and low scattering from bright flaring regions. In combination with the observations from the SOT and EIS, the XRT's broad temperature response, large dynamic range, and high throughput will achieve breakthrough science in the areas of CMEs (onset, coronal magnetic field structure, *etc.*), coronal heating (loop temperature dynamics, waves, and loop – loop interactions), flares, reconnection and jets (including the role of magnetic topology and energetics), and the relationship of the photospheric magnetic field evolution to coronal dynamics.

Acknowledgements We would like to thank Larry Hill and his staff at MSFC for effective and helpful management of this program and John Davis of MSFC for his experienced and useful advice and suggestions. We thank J. Pasachoff and the staff at Williams College for assistance with the VLI calibration. We also thank the staff at Goodrich (Danbury) for their effort and skill in fabricating the X-ray telescope and for permission to use the mirror performance figures.

References

Barbera, M., *et al.*: 2004, In: Hasinger, G., Turner, M.J.L. (eds.) *UV and Gamma-Ray Space Telescope Systems, Proc. SPIE* **5488**, 423.

Craig, I.J.D., Brown, J.C.: 1976, *Astron. Astrophys.* **49**, 239.

Deluca, E., *et al.*: 2005, *Adv. Space Res.* **36**, 1489.

Golub, L.: 2003, *Rev. Sci. Instrum.* **74**, 4583.

Kano, R., *et al.*: 2007, *Solar Phys.*, submitted.

Nariai, K.: 1987, *Appl. Opt.* **26**, 4428.

Nariai, K.: 1988, *Appl. Opt.* **27**, 345.

Smith, R.K., Brickhouse, N.S., Liedahl, D.A., Raymond, J.C.: 2001, *Astrophys. J.* **556**, L91.

Tsuneta, S., *et al.*: 1991, *Solar Phys.* **136**, 37.

Weber, M.A., Deluca, E.E., Golub, L., Sette, A.L.: 2005, In: Stepanov, A.V., Benevolenskaya, E.E., Kosovichev, A.G. (eds.) *Multi-Wavelength Investigations of Solar Activity, IAU Symp.* **223**, 321.

Werner, W.: 1977, *Appl. Opt.* **392**, 760.

The *Hinode* X-Ray Telescope (XRT): Camera Design, Performance and Operations

R. Kano · T. Sakao · H. Hara · S. Tsuneta · K. Matsuzaki · K. Kumagai · M. Shimojo · K. Minesugi · K. Shibasaki · E.E. DeLuca · L. Golub · J. Bookbinder · D. Caldwell · P. Cheimets · J. Cirtain · E. Dennis · T. Kent · M. Weber

Originally published in the journal Solar Physics, Volume 249, No 2.
DOI: 10.1007/s11207-007-9058-7 © Springer Science+Business Media B.V. 2007

Abstract The X-ray Telescope (XRT) aboard the *Hinode* satellite is a grazing incidence X-ray imager equipped with a 2048 × 2048 CCD. The XRT has 1 arcsec pixels with a wide field of view of 34 × 34 arcmin. It is sensitive to plasmas with a wide temperature range from < 1 to 30 MK, allowing us to obtain TRACE-like low-temperature images as well as *Yohkoh*/SXT-like high-temperature images. The spacecraft Mission Data Processor (MDP) controls the XRT through sequence tables with versatile autonomous functions such as exposure control, region-of-interest tracking, flare detection, and flare location identification. Data are compressed either with DPCM or JPEG, depending on the purpose. This results in higher cadence and/or wider field of view for a given telemetry bandwidth. With a focus adjust mechanism, a higher resolution of Gaussian focus may be available on-axis. This paper follows the first instrument paper for the XRT (Golub *et al.*, *Solar Phys.* **243**, 63, 2007) and discusses the design and measured performance of the X-ray CCD camera for the XRT and its control system with the MDP.

Keywords Sun: corona · Sun: X-rays

R. Kano (✉) · H. Hara · S. Tsuneta · K. Kumagai
National Astronomical Observatory of Japan, 2-21-1 Osawa, Mitaka, Tokyo 181-8588, Japan
e-mail: ryouhei.kano@nao.ac.jp

T. Sakao · K. Matsuzaki · K. Minesugi
Institute of Space and Astronautical Science, Japan Aerospace Exploration Agency, 3-1-1 Yoshinodai, Sagamihara, Kanagawa 229-8510, Japan

M. Shimojo · K. Shibasaki
Nobeyama Solar Radio Observatory, National Astronomical Observatory of Japan, Nobeyama, Nagano 384-1305, Japan

E.E. DeLuca · L. Golub · J. Bookbinder · D. Caldwell · P. Cheimets · J. Cirtain · E. Dennis · T. Kent · M. Weber
Smithsonian Astrophysical Observatory, 60 Garden Street, Cambridge, MA 02138, USA

1. Introduction

The *Solar-B* satellite was launched at 06:36 on 23 September 2006 (Japan Standard Time), and then named *Hinode* (Kosugi *et al.*, 2007). The X-Ray Telescope (XRT) aboard *Hinode* is a grazing incidence telescope designed to observe all the coronal features across a range of temperature not available to normal incidence telescopes. Golub *et al.* (2007) describes the XRT scientific objectives and requirements and the design and performance of the telescope portion of XRT. This companion paper describes the X-ray camera.

The XRT has the broadest temperature coverage of any coronal imager to date and an unprecedented combination of spatial resolution and field of view (Golub *et al.*, 2007). In the conceptual design phase of the strawman *Solar-B* mission around 1994 – 1996, there was an intense discussion on whether we should choose grazing incidence optics or normal incidence optics. Normal incidence optics can bring better spatial resolution, but a grazing incidence telescope is essentially sensitive to all temperatures. A dominant factor leading to the decision was the wide temperature range that the grazing incidence optics brings us. The great success of the TRACE mission undoubtedly shows the critical importance of high-resolution imaging (Golub *et al.*, 1999; Schrijver *et al.*, 1999). We, however, chose a balanced approach with regard to the conflicting scientific interests of spatial resolution, temperature range, and field of view. The choice resulted in apparent compromise in the spatial resolution with the obvious advantage of the wider field of view. A large-format 2048×2048 CCD allows coverage of the whole Sun, when the Sun is located at the center of the CCD. Note, however, that in nominal observing mode, *Hinode* tracks active regions or specific targets on the Sun that the other two telescopes aboard *Hinode* – the Solar Optical Telescope (SOT; Tsuneta *et al.*, 2007; Suematsu *et al.*, 2007; Ichimoto *et al.*, 2007; Shimizu *et al.*, 2007; Tarbell *et al.*, 2007) and the EUV Imaging Spectrometer (EIS; Culhane *et al.*, 2007) – also want to see, because they have a much smaller field of view as compared to the XRT.

The spatial resolution of a telescope depends on the combination of focal length and pixel size (as well as quality of the mirror, of course). A longer focal length and a smaller pixel size make a higher spatial resolution. The lowest limit of the pixel size is, however, given by the requirement on the dynamic range of the image, because a too small pixel size causes a too small full-well capacity. The length of a grazing incidence telescope is almost identical to its focal length. Therefore, to improve the spatial resolution of the XRT, we designed a longer telescope. A heroic effort was made by the spacecraft design team: Namely, the whole spacecraft design was driven to make the XRT (and EIS) as long as possible. Figure 1 of the *Hinode* outlook shows that the XRT is as long as the entire length of the spacecraft. This results in a pixel size of one arcsec.

Figure 2 indicates the four functional components constituting the XRT. XRT-T consists of the telescope metering tube including the X-ray and visible-light optics and focal-plane mechanisms (filter wheels, shutter, and focus motor). XRT-D is the driving electronics for the focal plane mechanisms. XRT-S is the focal-plane CCD imager attached at the rear end of XRT-T. It includes a 2048×2048 CCD device. XRT-E is the CCD electronics. The space-craft Mission Data Processor (MDP) also plays a vital role for XRT. The *Hinode* XRT is the result of the Japan – United States collaboration involving the Smithsonian Astrophysical Observatory (SAO) with NASA MSFC (XRT-T and XRT-D), the Institute of Space and Astronautical Science, the Japan Aerospace Exploration Agency, and the National Astro-nomical Observatory of Japan (XRT-S, XRT-E, and MDP). In Section 2, we discuss in detail the design, calibration, and measured performance of the CCD camera (XRT-S and XRT-E; Figure 3). Section 3 describes the observation control of the XRT with the MDP, and brief description of the camera thermal performance is given in Section 4.

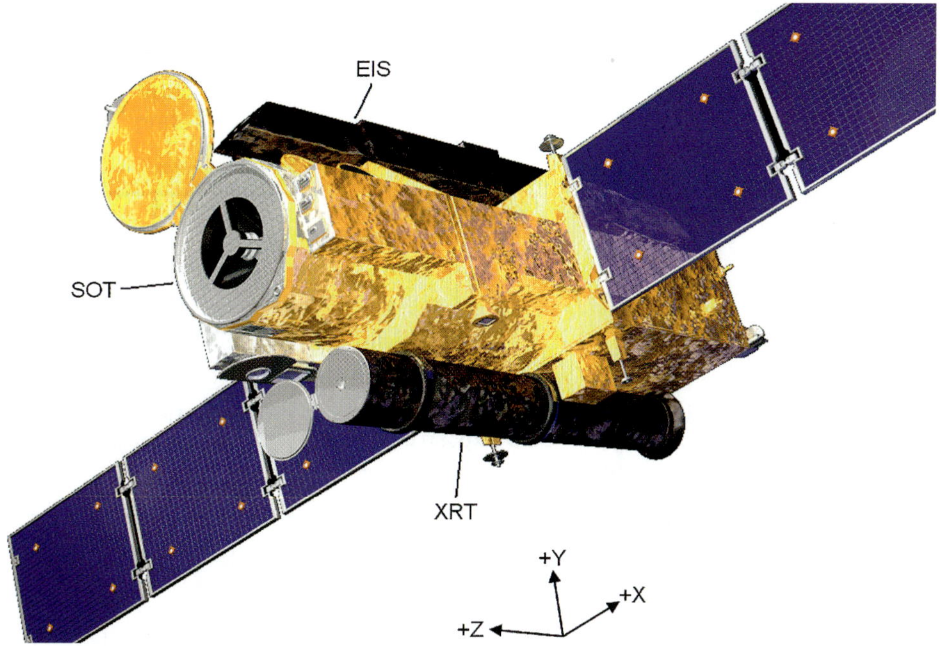

Figure 1   The *Hinode* satellite. The XRT is located on the $-Y$ side of the satellite body.

2. Instruments

Grazing incidence optics has a wide and continuous wavelength coverage. The XRT utilizes a back-illuminated CCD to enhance the sensitivity of longer wavelengths as compared with *Yohkoh*/SXT (Tsuneta *et al.*, 1991), which was equipped with a front-illuminated CCD. This essentially makes the XRT TRACE-like when we image the Sun with thin filters and SXT-like with thick filters as far as temperature sensitivity is concerned. What we will see with this telescope is a mixture of these aspects, depending on the differential emission measure within a pixel.

2.1. Camera Structure

XRT-S consists of a flat cylinder of the camera chassis and a skirt to the camera radiator (Figure 4). At the center of the chassis, the CCD housing is supported by two sets of thin flexures. This is a unique feature of the XRT that provides a focus adjustment with a stroke of ± 1 mm along the optical axis by commands from the ground. This design was driven by consideration of risk mitigation, which significantly reduces the risk of defocus in orbit, which would otherwise be difficult to overcome given the size and complexity of the X-ray telescope. In addition to risk mitigation, the focus adjustment capability can be used in scientific operation. The image plane of the grazing incidence optics is heavily curved, as shown in Figure 3 of Golub *et al.* (2007). The focus adjustment allows us to choose either on-axis maximum resolution with rapid off-axis degradation (Gaussian focus; see the curve for 2700.00 mm in the figure) or the focus position that gives resolution as uniform as possible over a larger field of view (optimized focus; see the curve for 2699.77 mm). Figure 5 shows the linkage structure of the focus rods in XRT-S. At points A and D, the focus rod system is

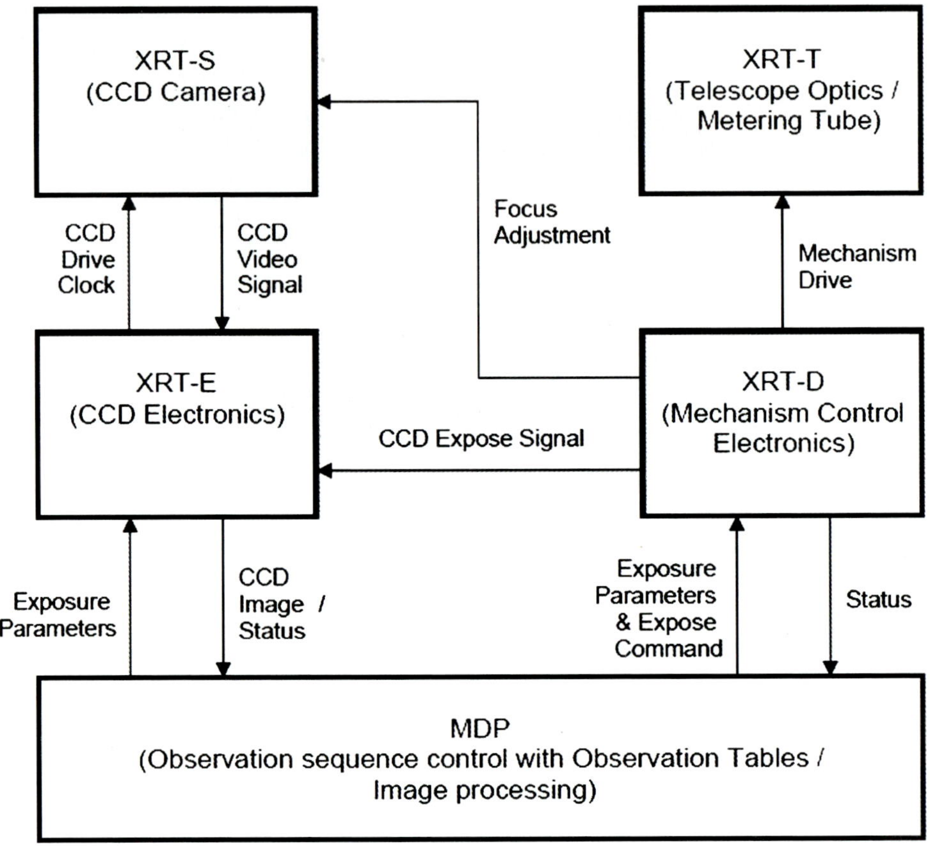

Figure 2 Block diagram showing four functional components of the XRT together with the MDP.

fixed to the chassis. Elastic hinges at A – E and an elastic pin at F give flexibility to the rod system. A push – pull motion from the focus motor mounted in XRT-T is transmitted to the portion F, and then the CCD housing is pushed or pulled from behind ($-Z$ side).

The housing, which is made of Invar, is thermally isolated from the surroundings by the thin flexures and the gold plating on them and connected to the camera radiator with two flexible thermal straps (left of Figure 4). The radiator always points away from the Sun and is designed to keep the CCD housing and the CCD in it at a temperature below $-43°$C. The thermal capacities of the housing and the radiator reduce the orbital variation of the CCD temperature. The Invar housing also shields the CCD against high-energy particles, to avoid degradation of the image quality.

The major components for XRT-S were extensively baked and outgases from them were measured with TQCMs during the assembly of XRT-S. For the outgas measurement, the TQCM temperature was set to $-84°$C to $-88°$C, because the coldest on-orbit (and also during pre-launch tests) predicted temperature for the CCD was $-78°$C, whereas the components were at room temperature. We set a mass accumulation rate of 1.56×10^{-9} g cm^{-2} h^{-1} onto the TQCM as a goal, which corresponds to an accumulation thickness of 120 Å/month in the case of material with $\rho \approx 1$ g cm^{-3}. Although some components did not achieve the goal even after baking, we comfortably accepted them (in the worst case, the mass accumu-

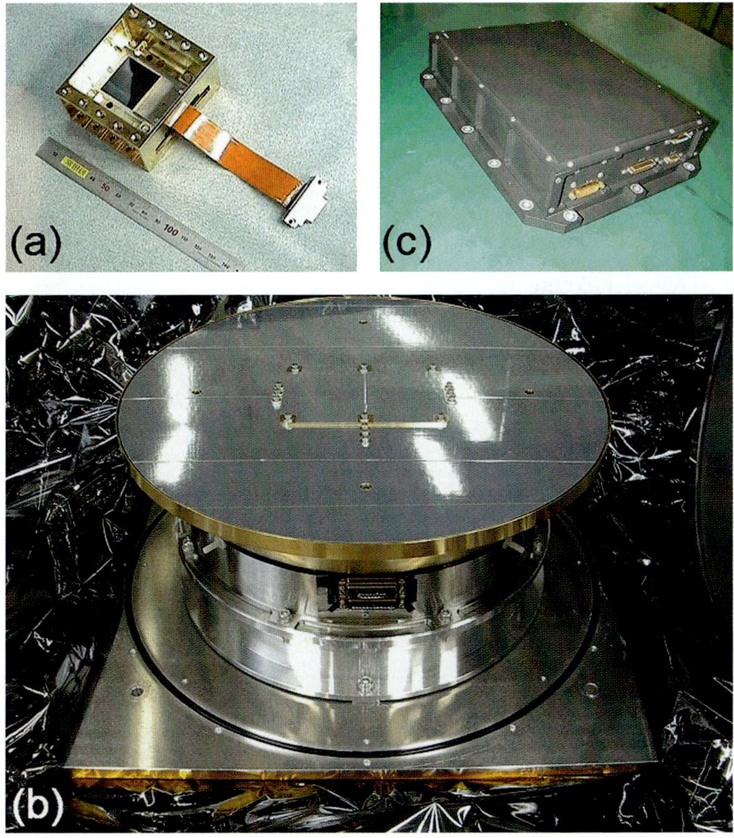

Figure 3   (a) The XRT CCD installed in the CCD Invar housing (a mechanical sample). (b) XRT-S, the XRT camera, mounted on the base plate of its shipping container. (c) XRT-E, the camera electronics.

lation rate was 5×10^{-9} g cm^{-2} h^{-1}), because they were to be used outside of the camera. On the rear side of the CCD package, an on-orbit decontamination heater is attached. It will be turned on to avoid the accumulation of outgases on the CCD in the initial phase and to remove the accumulated outgases in the later phase.

2.2. CCD Performance

The XRT uses a back-illuminated three-phase CCD with 13.5-μm pixel size and a 2048 × 2048 array, which was manufactured by e2v Technologies. The CCD has two identical readout ports: an R-port and an L-port. The XRT uses the R-port as the default port and the L-port as a backup. From either port, an entire CCD image can be read. Camera performance (camera gain, dark current, quantum efficiency, and CCD cooling) was calibrated in the Advanced Technology Center of the National Astronomical Observatory of Japan. Basic features of the camera, including some results from the calibration are briefly summarized in Table 1. The camera system gain, the dark current, and the quantum efficiency are described in detail in the subsequent sections.

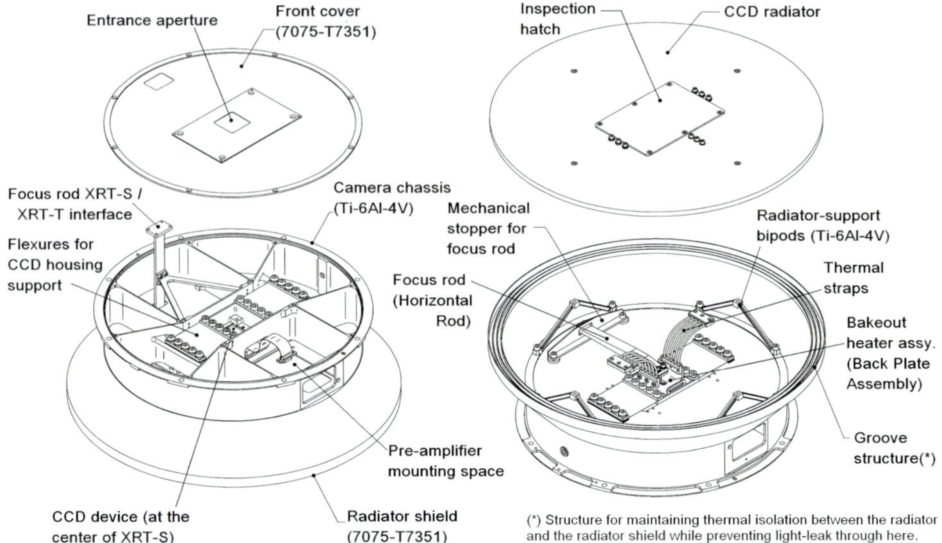

Figure 4 Schematic drawings of XRT-S. Left: Seen from the telescope side (Sun-facing side) with the front cover detached. Right: From the rear side (radiator side) with the radiator detached. Note that the pre-amplifier, which is mounted inside XRT-S, is not shown in the figure.

Table 1 XRT camera characteristics.

CCD type	Back-illuminated three-phase CCD (e2v/CCD 42-40)
Pixel format	2048 × 2048 pixels
Pixel size	13.5 × 13.5 μm (corresponding to 1 × 1 arcsec)
Field of view	34 × 34 arcmin
Pixel binning mode	1 × 1, 2 × 2, 4 × 4 and 8 × 8
Dark current	0.1 e^-/s/pixel @ $-65°$C
CCD temperature	Passive cooling: $< -43°$C
CTE	Parallel: > 0.999996,
	Serial: > 0.999999 ($-93°$C $< T < -50°$C)
QE (X-ray/EUV)	0.93 @ 13 Å, 0.61 @ 45 Å, 0.46 @ 114 Å, 0.56 @ 304 Å
QE (visible light)	0.44 @ 4000 Å, 0.66 @ 5000 Å
Full-well capacity	2.0×10^5 e^-
Camera gain constant	57 e^-/DN
Camera system noise	$< 30 e^-$
Output data resolution	12 bit

2.2.1. Camera Gain

The XRT camera gain was measured with an ^{55}Fe isotope at several CCD temperatures. Figure 6 (top) shows an example of the ^{55}Fe isotope data. Mn-Kα and Mn-Kβ lines are clearly seen. Figure 6 (bottom) summarizes the measured camera gain for both read-out ports over the CCD operation temperature from $-43°$C to $-100°$C. Although there is a slight dependence on the CCD temperature as shown in the following, we can reasonably adopt a value of 57 e^-/DN as the typical value for this temperature range. The fitted lines

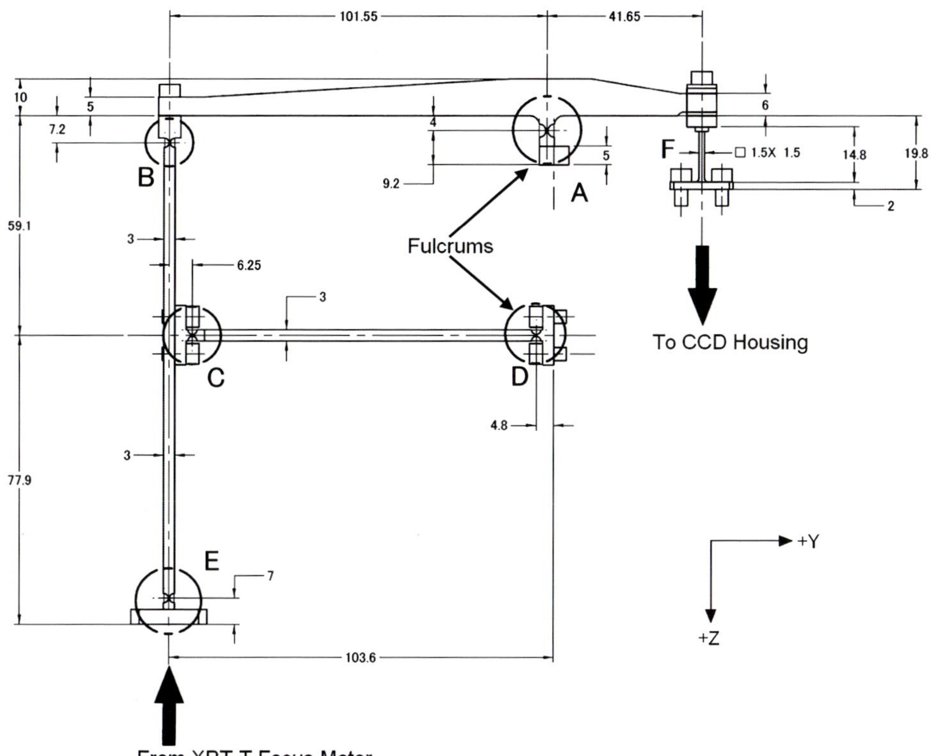

Figure 5 Focus mechanism rod system inside XRT-S. Portions shown by circles with labels A – E are elastic hinges. A and D are fixed to the XRT-S chassis, serving as fulcrums. The spacecraft coordinates are also shown.

are

$$G_R \ [e^-/\text{DN}] = 59.1 + 0.026\, T \ [^\circ\text{C}], \tag{1}$$

$$G_L \ [e^-/\text{DN}] = 58.8 + 0.034\, T \ [^\circ\text{C}]. \tag{2}$$

2.2.2. Dark Current

Figure 7 shows the temperature dependence of the dark current and three model curves calculated by the following general formula (Janesick, 2000):

$$D_R \ [e^-/\text{s/pixel}] = 2.5 \times 10^{15} \ P_S \ D_{FM} \ T^{1.5} \ \exp[-E_g/(2kT)], \tag{3}$$

where P_S is the pixel area (cm^2), D_{FM} is called the "dark current figure of merit" at 300 K (nA cm^{-2}), T is the CCD temperature (K), k is Boltzmann's constant (8.62×10^{-5} eV K^{-1}), and E_g is the bandgap energy (eV) described by the following empirical formula:

$$E_g \ [\text{eV}] = 1.1557 - \frac{7.021 \times 10^{-4} T^2}{1108 + T}. \tag{4}$$

Below -75°C, the dark current was too small to derive meaningful values with the longest available exposure (64 seconds) with the test setup for the calibration. The remaining data

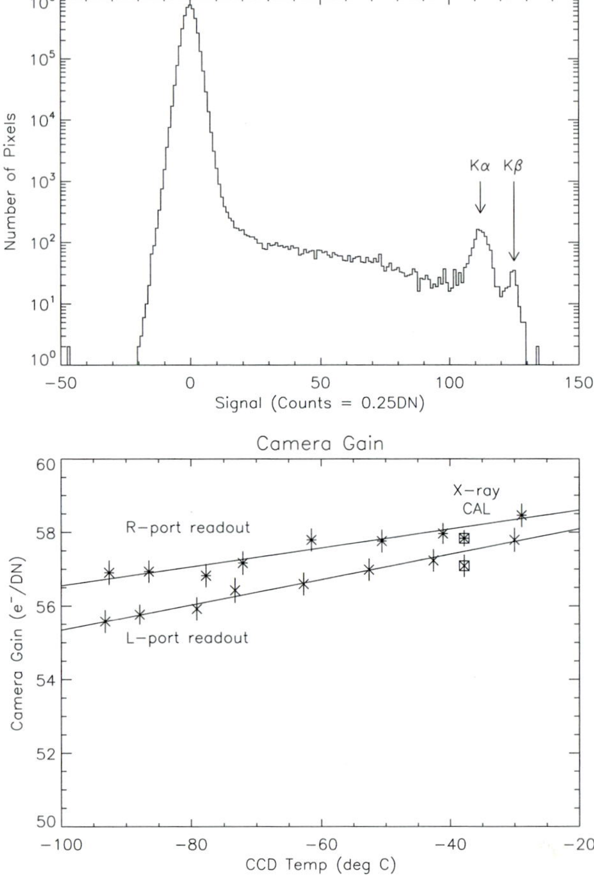

Figure 6 (Top) A histogram showing signal distribution for an ^{55}Fe image taken through the R-port at $-93°$C, with dark current subtracted. Mn-Kα and Mn-Kβ lines are clearly seen around the signal values of 110 and 125, respectively. (Bottom) The system camera gain of the XRT camera taken during the thermal vacuum test (crosses). Error bars show 3σ statistical uncertainty. The data separately taken during the X-ray QE measurements are also shown by boxes.

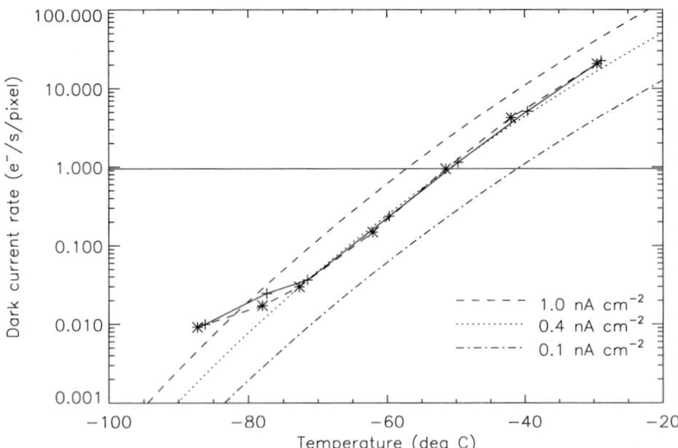

Figure 7 Temperature dependence of dark current for the XRT CCD. Crosses and stars show the data from the R-port and the L-port, respectively. The horizontal line shows a level of 1 DN/64 seconds. The dark current below $-75°$C is overestimated and should not be regarded as real.

points are well fitted to the model curve with $D_{\mathrm{FM}} = 0.4\ \mathrm{nA\ cm^{-2}}$. Because the CCD will be operated below $-43°$C on orbit, the dark current is at most a few DN even with the longest exposure for the camera (64 seconds).

Table 2 Emission lines used for X-ray and EUV QE measurements.

Line	Wavelength (Å)	Line	Wavelength (Å)	Line	Wavelength (Å)
Mn K	2.1[a]	Ni L	14.6	Be K	114
Mo L	5.4	Fe L	17.6[a]	He II	256
W K	7.0	O K	23.6	He II	304[a]
Al K	8.3[a]	Ti L	27.4	He I	584
Mg K	10.0[a]	C K	44.6[a]		
Cu L	13.3	Mo M	64.2		

[a]L-port read-out data were only taken at six wavelengths.

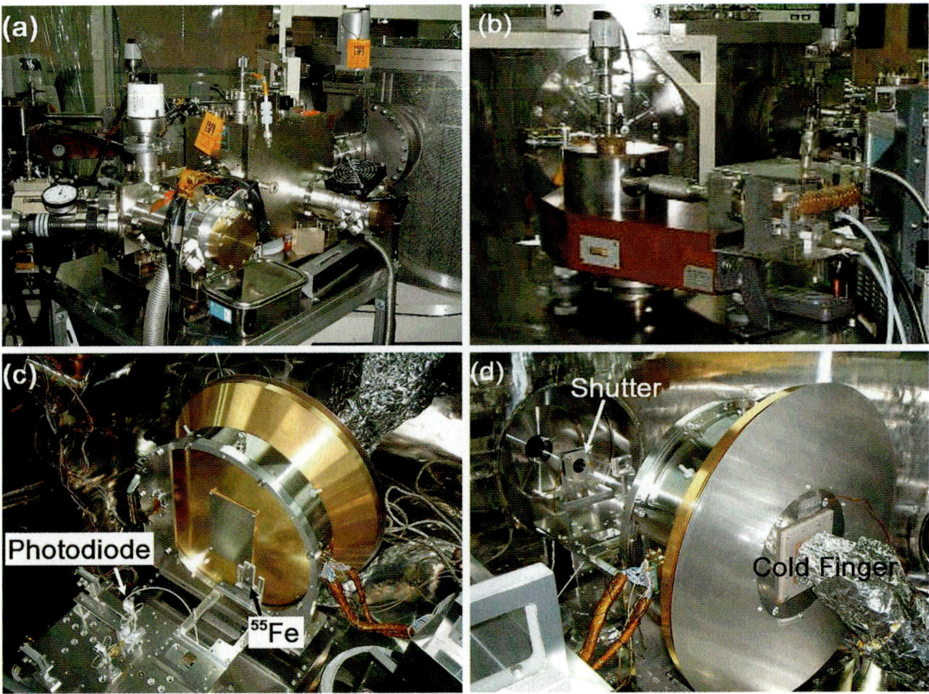

Figure 8 QE measurement configurations. Panels (a) and (b) show X-ray and EUV monochromators, respectively. Panels (c) and (d) show the setup in the calibration chamber with XRT-S. A reference photodiode, an ^{55}Fe isotope, and a beam shutter were located in front of XRT-S. A cold finger was attached to the camera radiator from behind.

2.2.3. *Quantum Efficiency*

The X-ray and EUV quantum efficiency was measured with 16 emission lines listed in Table 2. From Mo-L 5.4 Å to Be-K 114 Å, we used the in-focus monochromator (IFM-SXR-0.5) manufactured by Hettrick Scientific, Inc., with the Manson Model 2 X-ray Source of Austin Instruments, Inc. (Figure 8a). Above He-II 256 Å, we used the EUV monochromator (LHT30) of Jobin-Yvon (Figure 8b). Figures 8c and 8d show the experiment configuration in

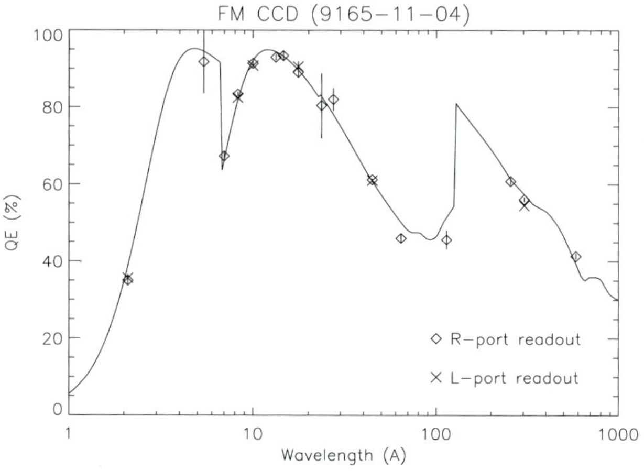

Figure 9 The quantum efficiency of the XRT CCD. Error bars shows 1σ statistical uncertainty. The solid curve indicates the best-fit QE curve from a CCD model described in the text.

the calibration chamber. A silicon photodiode manufactured by International Radiation Detectors (IRD) was used as the reference detector for the XRT CCD. The photodiode was calibrated by NIST (National Institute of Standards and Technology) in the wavelength range between 50 and 1200 Å. We also received information on the thicknesses of the depletion layer and the oxide layer of this photodiode from the manufacturer, and we extrapolated the photodiode quantum efficiency (QE) with a simple model assuming a silicon thickness of 25 μm with 120-Å top oxide layer and a unity charge collection efficiency in the silicon. For Mn-K 2.1 Å, we used an ^{55}Fe isotope whose X-ray count rate was calibrated by ourselves with the Manson Model 04 Gas Flow Proportional Counter.

Figure 9 shows the wavelength dependence of the quantum efficiency of the XRT CCD, with the best-fit QE curve. As shown in Stern, Shing, and Blouke (1994), the charge collection efficiency (η) of back-illuminated CCD is low near the back surface. We adopted a simplified expression for this effect with an exponential function,

$$\eta(x) = 1 - (1 - \eta_0)\exp(-\gamma x),\tag{5}$$

instead of a linear function in Stern, Shing, and Blouke (1994) of

$$\eta(x) = \begin{cases} \eta_0 + (1 - \eta_0)\gamma x & \text{for } 0 < x < \gamma^{-1}, \\ 1 & \text{for } x > \gamma^{-1}, \end{cases}\tag{6}$$

to avoid the somewhat clear boundary at a depth of γ^{-1} in the latter formula. In these equations, η_0 is the charge collection efficiency at the back surface, γ is the inverse of the thickness of the back-surface effect, and x is the distance from the CCD back surface. Following Stern, Shing, and Blouke (1994), we derive a model QE curve as

$$QE = e^{-\mu_{\text{SiO}_2} d_{\text{SiO}_2}} \int_0^{d_{\text{Si}}} \eta(x)\mu_{\text{Si}} e^{-\mu_{\text{Si}} x} dx \tag{7}$$

$$= e^{-\mu_{\text{SiO}_2} d_{\text{SiO}_2}} \left\{ 1 - e^{-\mu_{\text{Si}} d_{\text{Si}}} - (1 - \eta_0)\frac{\mu_{\text{Si}}}{\mu_{\text{Si}} + \gamma}\left(1 - e^{-(\mu_{\text{Si}} + \gamma)d_{\text{Si}}}\right) \right\},\tag{8}$$

where d_{Si} is the thickness of the silicon substrate, d_{SiO_2} is the thickness of the oxidized layer on the back surface, and μ_{Si} and μ_{SiO_2} are the absorption coefficients of silicon and silicon

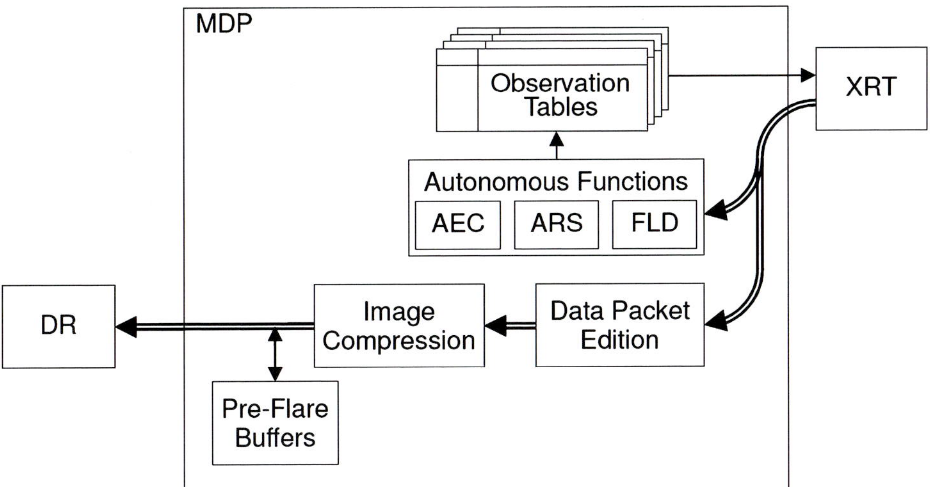

Figure 10 Functional block diagram of the MDP for XRT observations. Arrows with double lines show the flow of image data. All the XRT images pass through the data packet edition and image compression and are stored in the data recorder (DR). Some of the images are also transferred to the functional blocks for autonomous functions where they are analyzed to update the observation parameters in the MDP.

dioxide, respectively. The best-fit curve shown in Figure 9 is calculated with parameters of $d_{\mathrm{Si}} = 14.4$ μm, $d_{\mathrm{SiO_2}} = 66$ Å, $\eta_0 = 0.41$, and $\gamma = (0.21\ \mu\mathrm{m})^{-1}$.

3. Observation Control System

The XRT is scientifically controlled by the spacecraft MDP (see Figure 3.1). The MDP can perform onboard processing of the XRT image data for autonomous observation control such as selection of regions of interest (Automatic Region Selector; ARS), Automatic Exposure Control (AEC), and Flare Detection (FLD). The MDP also has dedicated pre-flare buffers within which series of pre-flare images can be stored. In this section, we describe the MDP functions for XRT observations from the viewpoint of an observer preparing his or her proposed observations.

3.1. Observation Tables

The MDP controls the XRT based on the observation tables stored inside it. The observation tables consist of three hierarchies of tables: main programs, subroutines, and sequence tables.

Each sequence table includes a maximum of eight exposure commands and is prepared as an elemental set of the XRT observations. For example, one sequence table may be designed for the full-disk temperature analysis with two alternate filters, and another for the DEM analysis with eight different filters in succession.

The sequence tables are building blocks for the main programs and subroutines. The observer can prepare the main programs and subroutines by combining the existing or newly prepared sequence tables for his or her proposed observations. The observer can also prepare a different main program for flare observations that is activated when the flare flag is set by the Flare Detection logic.

Table 3 ROI table stored in the MDP.

ROI No.	Purpose
1	The brightest region detected by the ARS global search.
2 – 4	Bright regions tracked by the ARS local search.
5 – 12	Static (not updated by MDP) ROIs set by the observer. (Note: ROI5 is also used for the Pre-Flare function.)
15 – 16	Flares detected by the FLD function.

3.2. Region-of-Interest Management

The XRT can take partial frame images. The horizontal and vertical size of the partial frame images are 64, 128, 192, 256, 384, 512, 768, 1024, 1536, and 2048 pixels. The observer can select either square or rectangular-shaped partial frame images (*e.g.*, 1024×256, 256×512). The smallest size, 64 pixels, is only available in a square image format. If the spacecraft is pointed at the center of the solar disk, the largest field of view (2048×2048 pixel = 34×34 arcmin) can cover the full solar disk.

In the MDP, up to 16 region-of-interests (ROIs) for observations can be managed with a table dedicated for them. Each ROI is specified by its location and size on the XRT CCD and is given a unique ID number. Each exposure command refers to the ROI by its ID number in a sequence table. The observer can set the locations and sizes for the ROIs before observation. Among 16 ROIs, the locations of ROI1 – ROI4 and the locations of ROI15 and ROI16 are dynamically updated by the ARS function and by the FLD function of the MDP, respectively (Table 3), during the observation as described later. ROI sizes are not updated by the MDP and are maintained as the values set by the observer.

3.3. Automatic Exposure Control

Although X-ray luminosity of the corona ranges over eight orders of magnitude from coronal holes and outer corona to intense X-class flares, the CCD has a dynamic range of only three orders of magnitude. Often the coronal intensity changes rapidly when flares start. The correct exposure is realized by changing the exposure duration over a wide range. A set of available exposure durations is shown in Table 4. Table 5 shows the typical count rate for various coronal features through different filters. The MDP analyzes X-ray images onboard right after each exposure, and adjusts their exposure duration in pipeline manner. This function is called Automatic Exposure Control (AEC). [Note that AEC is available for images whose size is smaller than or equal to 262 144 pixels (*i.e.*, 512×512 pixel image or 2048×128 pixel image).] If an X-ray image does not achieve the proper exposure with the shortest exposure, AEC automatically changes the X-ray analysis filter to a thicker filter prespecified by the observer in sequence tables.

Note that AEC can be disabled for any exposure in observation tables. It is also possible to intentionally take overexposed or underexposed images for any exposure by changing the AEC parameters. This is often quite useful as demonstrated by *Yohkoh*, for observing, *e.g.*, faint objects surrounding a bright structure.

3.4. Automatic Region Selector

The Automatic Region Selector (ARS) is the function that is used to search for bright regions and to automatically change the coordinates of the partial frame images to observe the new

Table 4 Exposure table.

Index : Exposure	Index : Exposure	Index : Exposure	Index : Exposure
0 : 1 ms	8 : 24 ms	16 : 354 ms	24 : 5.66 s
1 : 2 ms	9 : 32 ms	17 : 500 ms	25 : 8.00 s
2 : 3 ms	10 : 44 ms	18 : 707 ms	26 : 11.3 s
3 : 4 ms	11 : 63 ms	19 : 1.00 s	27 : 16.0 s
4 : 5 ms	12 : 86 ms	20 : 1.41 s	28 : 22.6 s
5 : 8 ms	13 : 125 ms	21 : 2.00 s	29 : 32.0 s
6 : 12 ms	14 : 177 ms	22 : 2.83 s	30 : 45.2 s
7 : 16 ms	15 : 250 ms	23 : 4.00 s	31 : 64.0 s

Table 5 Typical count rate (in units of DN/pixel).

Exposure (s)	Coronal hole		Quiet Sun		Active region		Flare (M2)	
	45.2	64.0	11.3	32.0	0.50	4.0	0.001	0.004
Thin-Al/Mesh	124	175	300	850	949	sat.	sat.	sat.
Thin-Al/Poly	16	23	139	394	894	sat.	sat.	sat.
C/Poly	3	5	74	209	544	sat.	sat.	sat.
Ti/Poly	3	5	57	161	386	sat.	sat.	sat.
Thin-Be	0	0	15	44	204	1632	sat.	sat.
Med-Be	0	0	2	5	27	219	1419	sat.
Med-Al	0	0	1	3	14	111	619	2477
Thick-Al	0	0	0	0	2	14	90	361
Thick-Be	0	0	0	0	0	0	6	23

References of the model DEMs: Data for "Coronal hole," "Quiet Sun," and "Active region" are adopted from Vernazza and Reeves (1978), and "Flare (M2)" data are from Dere and Cook (1979). If the count rate exceeds 3000 DN, it is shown as "sat."

region. For this purpose, the XRT takes full-frame CCD images with a 2-arcsec resolution (ARS patrol images) at a regular interval. The time resolution of ARS (*i.e.,* update interval for the region selection) depends on the cadence of ARS patrol images that can be set in the table. The baseline of the ARS time resolution is about 1.5 h.

There are global search and local search modes in ARS, which function independently of each other. The global search selects the brightest region on the entire XRT CCD and updates the location of ROI1 (see Table 3). The local search tracks the brightest region in each of ROI2, 3, and 4, by searching a limited area around the current location. The locations and sizes for these ROIs are initially set by the observer. With the local search we can track up to three targets in parallel.

3.5. Flare Detection

Hinode has no independent X-ray detection system dedicated to identify solar flares. Thus, XRT has to do this by itself using a Flare Detection (FLD) algorithm. FLD automatically identifies the occurrence of a flare, then determines the flare location on the CCD, and finally sets a flare flag for the XRT as well as for the SOT and the EIS. For this purpose, XRT takes full-frame CCD images with an 8-arcsec resolution (called FLD patrol images) at a regular interval. The baseline of FLD patrol interval is about 30 s.

The method to identify flares is not based on a simple intensity threshold monitoring of the FLD patrol images. Because many solar flares are generally not so bright at the beginning in soft X rays, if flares are to be detected as soon as they occur, a simple threshold-based detection algorithm does not suffice. A better FLD algorithm is to monitor the increase in intensity by comparing with a running-averaged patrol image generated based on a collection of FLD patrol images taken previously. The MDP calculates the parameter q^2, which is actually a map to represent the increase of the X-ray intensity normalized by photon noise:

$$q^2(x, y) = \frac{[F(x, y) - F_{\mathrm{avg}}^{(i-1)}(x, y)]^2}{F_{\mathrm{avg}}^{(i-1)}(x, y) + g} \quad \text{for } F > F_{\mathrm{avg}}^{(i-1)}, \tag{9}$$

where g is a control parameter to avoid division by 0, F is the patrol image to be evaluated, and $F_{\mathrm{avg}}^{(i)}$ is the running-averaged patrol image calculated by

$$F_{\mathrm{avg}}^{(i)}(x, y) = \gamma F(x, y) + (1 - \gamma) F_{\mathrm{avg}}^{(i-1)}(x, y). \tag{10}$$

γ is also a parameter that controls the effective duration of the running average. If q^2 exceeds a threshold for flare start, the MDP sets the flare flag and finds a flare location in the F-map around the peak location in the q^2-map. The flare location will be dynamically set as the center position of ROI15 and ROI16, which are then referred to from sequence tables. The observer can set the sizes of ROI15 and ROI16 before observations. When q^2 becomes lower than a threshold for flare end, the MDP drops the flare flag.

FLD can also detect radiation belts. In radiation belts, the MDP changes the parameters of the flare detection algorithm to avoid spurious effects of charged particles in identifying flare occurrence.

3.6. Image Compression

The CCD video signal is digitized with a 14-bit analog-to-digital converter in XRT-E, the upper 12 bits of which are then sent to the MDP. The observer can specify three types of compression for each exposures in sequence tables: no compression, lossless compression, and lossy compression. In the no-compression mode, the MDP is transparent. In lossless compression, the MDP compresses images with the differential pulse code modulation (DPCM) method, which does not lose any information of the image. The efficiency of DPCM for XRT images is expected to be about 50% (*i.e.,* 6 bits/pixel) according to simulation studies. This is a popular option to reduce the amount of telemetry data. In the lossy compression, the MDP compresses images with the JPEG method, whose quality factor can be specified for each exposure when preparing observation tables depending on the purpose. JPEG compression may be useful for providing context images for SOT and EIS or for performing purely morphological studies.

3.7. Exposure Cadence

The time cadence for a series of exposures depends on the following pre- and post-exposure activities: (1) setup time for the focal-plane mechanisms, especially the filter wheel movements (a movement to an adjacent filter position takes about 0.8 seconds); (2) read-out time of the CCD image (9.3 seconds for a full-frame CCD image); (3) image processing time

Table 6 Time cadences for three typical examples.

Example 1	Continuous observation of an active region
ROI	FOV = 384 × 384 arcsec, Binning = 1 × 1 arcsec
Time interval of a pair of filter images	
	30 s
Data rate	590 000 pixels min^{-1}
Example 2	High-speed observation of an active region
ROI	FOV = 384 × 384 arcsec, Binning = 1 × 1 arcsec
Time interval of a pair of filter images	
	5 s
Period	10 min observation and 50 min intermission
Data rate	590 000 pixels min^{-1}
Example 3	Combination of full frame and partial frame images
ROI1	FOV = 384 × 384 arcsec, Binning = 1 × 1 arcsec
ROI2	FOV = 2048 × 2048 arcsec, Binning = 4 × 4 arcsec
Time interval of a pair of filter images	
	ROI1 = 40 s and ROI2 = 200 s
Data rate	600 000 pixels min^{-1}

in the MDP (with autonomous functions in the MDP lagging the exposure cadence by as little as 2 seconds); and (4) restriction of the allocated data rate (as described later). If the observer plans to take images at a high rate, the number of filters and/or the image size may have to be adjusted to keep within the data rate allocated for the XRT.

The total data rate of the *Hinode* satellite depends on the frequency of the data recorder playback through downlink stations. A typical frequency of playback is about 15 downlinks/day. The total data rate is estimated to be 400 kbps for the scientific instruments. SOT, XRT, and EIS are typically allocated 70%, 15%, and 15% of the bandwidth, respectively. The XRT image data are 12 bits/pixel and would be compressed to about 6 bits/pixel by DPCM. Therefore, the typical data rate for the XRT is about 600 000 pixels min^{-1}.

The following three examples are shown in Table 6. (1) For a continuous observation of an active region, a pair of two filter images is taken every 30 seconds. (2) For a fast cadence observation of an active region, a pair of two filter images is taken every 5 seconds. Such a burst observation is available for a limited time interval to keep the average rate at about 600 000 pixels min^{-1}. (3) For a combination of observations of an active region and its surrounding region, a pair of two filter images covering an active region is taken every 40 seconds, while the same pair of images covering the entire CCD is taken every 200 seconds.

3.8. Pre-Flare Buffers

In the MDP, four memory buffers are prepared for XRT images. One of them is used for the normal image transfer. The other three are used as ring buffers dedicated for pre-flare images. The observer can select the use of these special buffers for pre-flare observations. Once the MDP detects the occurrence of a flare and if the flare occurs within the field of view of ROI5, MDP protects the data in the pre-flare buffers from being overwritten until the data are downlinked or the protection is released by a command. If the MDP does not detect a flare in ROI5, it keeps overwriting the previous data in the pre-flare buffers. Because ROI5 is used in the logic to freeze the pre-flare buffers, it is recommended that ROI5 be used for

Table 7 An example of a pre-flare observation.

ROI	FOV $= 384 \times 384$ arcsec, Binning $= 1 \times 1$ arcsec
Time interval of a pair of filter images	
Before $X - 160$ s	40 s ("X" is the onset time of a flare)
$X - 160$ s to $X - 100$ s	20 s
$X - 100$ s to X	10 s

pre-flare observations. The size of the pre-flare buffers is 5.5 Mbytes in total. The observer can compose a wide variety of pre-flare observation using these pre-flare buffers. Table 7 shows an example for the pre-flare observation that takes an active region image with a pair of X-ray analysis filters for the temperature diagnostics.

4. Thermal Performance of the CCD Camera

Just after launch, the decontamination heater for the CCD was turned on to avoid accumulation of outgas onto the CCD. During this CCD bakeout period, the heater had kept the CCD at about 30°C, while its surroundings were at 10°C. The radiator temperature was -40°C. One month later, the heater was turned off, and the radiator and CCD were cooled down to -75°C and -69°C, respectively. The orbital variation of the CCD temperature is smaller than ± 1°C during the noneclipsing period.

5. Conclusion

The initial observations of the XRT have already begun. The on-orbit performance of the XRT camera is excellent and has met or exceeded all pre-launch expectations. The XRT is proving to be a powerful tool for investigating the many forms of coronal activity and is expected to reveal, though coordinated observations with the *Hinode* Solar Optical Telescope and EUV Imaging Spectrometer, the mechanisms of coronal heating.

Acknowledgements We would like to thank our scientific and engineering colleagues of XRT at the Smithsonian Astrophysical Observatory (SAO), the National Astronomical Observatory of Japan (NAOJ), and the Institute of Space and Astronautical Science of JAXA (ISAS/JAXA). We also would like to thank Kenji Hiyoshi, Michihiro Horii, and Koji Taguchi of Meisei Electric Co., Ltd., for developing the XRT camera electronics, Satoru Iwamura, Zhangong Du, and Mitsuhiko Nakano of Astro Research Corp. for developing the XRT camera structures and thermal design, Peter Pool and Wolfgang Suske of e2v Technologies (UK), Ltd., and Kiyoshi Tabata of Cornes Dodwell, Ltd., for providing the best-quality CCD suited to XRT, Masahiro Koyama of Mitsubishi Heavy Industries, Ltd., for developing the MDP hardware, Katsuya Yamamoto and Masayuki Nagase of Systems Engineering Consultants Co., Ltd., for developing the MDP software, Tomonori Tamura of NAOJ for supporting the baking and outgas measurements of XRT-S, Tetsuo Nishino and Norio Okada of the Advanced Technology Center (ATC) of NAOJ for developing equipment for the XRT-S calibrations, Akira Ohnishi of ISAS/JAXA for advising on the thermal design of XRT-S, and Kazuyuki Hirose of ISAS/JAXA for advising on the electric design of XRT-S and XRT-E. We would like to thank the National Astronomical Observatory of Japan for financial support for developing the calibration facility in the ATC for the XRT CCD camera system. We wish to express our sincere gratitude to the late Prof. Takeo Kosugi, former project manager of *Hinode* at ISAS, who passed away suddenly in November 2006. Without his leadership in the development of *Solar-B/Hinode*, this mission would have never been realized. Finally, we would like to thank Roger Hauck of SAO for developing the electrical interface between XRT-D and XRT-E with us. Unfortunately, he passed away before the launch of *Hinode*. We hope his soul rests in peace.

References

Culhane, J.L., Harra, L.K., James, A.M., Al-Janabi, K., Bradley, L.J., Chaudry, R.A., *et al.*: 2007, *Solar Phys.* **243**, 19.

Dere, K.P., Cook, J.W.: 1979, *Astrophys. J.* **229**, 772.

Golub, L., Bookbinder, J., DeLuca, E., Karovska, M., Warren, H., Schrijver, C.J., *et al.*: 1999, *Phys. Plasmas* **6**, 2205.

Golub, L., DeLuca, E., Austin, G., Bookbinder, J., Caldwell, D., Cheimets, P., *et al.*: 2007, *Solar Phys.* **243**, 63.

Ichimoto, K., Lites, B., Elmore, D., Suematsu, Y., Tsunete, S., Katsukawa, Y., *et al.*: 2007, *Solar Phys.*, submitted.

Janesick, J.R.: 2000, *Scientific Charge-Coupled Devices*, SPIE, Bellingham.

Kosugi, T., Matsuzaki, K., Sakao, T., Shimizu, T., Sone, Y., Tachikawa, S., *et al.*: 2007, *Solar Phys.* **243**, 3.

Schrijver, C.J., Title, A.M., Berger, T.E., Fletcher, L., Hurlburt, N.E., Nightingale, R.W., *et al.*: 1999, *Solar Phys.* **187**, 261.

Shimizu, T., Nagata, S., Tsuneta, S., Tarbell, T., Edwards, C., Shine, R., *et al.*: 2007, *Solar Phys.*, in press.

Stern, R.A., Shing, L., Blouke, M.M.: 1994, *Appl. Opt.* **33**, 2521.

Suematsu, Y., Tsuneta, S., Ichimoto, K., Shimizu, T., Otsubo, M., Katsukawa, Y., *et al.*: 2007, *Solar Phys.*, submitted.

Tarbell, T.D., *et al.*: 2007, *Solar Phys.*, submitted.

Tsuneta, S., Acton, L., Bruner, M., Lemen, J., Brown, W., Caravalho, R., *et al.*: 1991, *Solar Phys.* **136**, 37.

Tsuneta, S., Suematsu, Y., Ichimoto, K., Shimizu, T., Otsubo, M., Nagata, S., *et al.*: 2007, *Solar Phys.*, submitted.

Vernazza, J.E., Reeves, E.M.: 1978, *Astrophys. J. Suppl.* **37**, 485.

The EUV Imaging Spectrometer for Hinode

J.L. Culhane · L.K. Harra · A.M. James · K. Al-Janabi · L.J. Bradley ·
R.A. Chaudry · K. Rees · J.A. Tandy · P. Thomas · M.C.R. Whillock · B. Winter ·
G.A. Doschek · C.M. Korendyke · C.M. Brown · S. Myers · J. Mariska · J. Seely ·
J. Lang · B.J. Kent · B.M. Shaughnessy · P.R. Young · G.M. Simnett · C.M. Castelli ·
S. Mahmoud · H. Mapson-Menard · B.J. Probyn · R.J. Thomas · J. Davila · K. Dere ·
D. Windt · J. Shea · R. Hagood · R. Moye · H. Hara · T. Watanabe · K. Matsuzaki ·
T. Kosugi · V. Hansteen · Ø. Wikstol

Originally published in the journal Solar Physics, Volume 243, No 1.
DOI: 10.1007/s01007-007-0293-1 © Springer 2007

T. Kosugi deceased 2006 November 26.

J.L. Culhane (✉) · L.K. Harra · A.M. James · K. Al-Janabi · L.J. Bradley · R.A. Chaudry ·
K. Rees · J.A. Tandy · P. Thomas · M.C.R. Whillock · B. Winter
Mullard Space Science Laboratory, University College London, Holmbury St Mary, Dorking,
Surrey, RH5 6NT, UK
e-mail: jlc@mssl.ucl.ac.uk

G.A. Doschek · C.M. Korendyke · C.M. Brown · S. Myers · J. Mariska · J. Seely
Naval Research Laboratory, E.O. Hulburt Centre for Space Research, Washington, DC 20375-5320,
USA

J. Lang · B.J. Kent · B.M. Shaughnessy · P.R. Young
Space Science and Technology Department, Rutherford Appleton Laboratory, Chilton, Didcot,
Oxfordshire, OX11 0QX, UK

G.M. Simnett · C.M. Castelli · S. Mahmoud · H. Mapson-Menard · B.J. Probyn
Space Research Group, School of Physics and Space Research, University of Birmingham,
Birmingham, UK

R.J. Thomas · J. Davila
NASA Goddard Space Flight Centre, Code 682, Greenbelt, MD 20771, USA

K. Dere
School of Computational Sciences, George Mason University, 4400 University Drive, Fairfax,
VA 22030, USA

D. Windt
Pupin Physics Laboratories, Department of Astronomy, Columbia University, 550 West 120th
Street, New York, 10027, USA

J. Shea
Perdix Corporation, P.O. Box 23, 35 Howard Street, Wilton, NH 03086, USA

R. Hagood
Swales Aerospace, 5050 Powder Mill Road, Beltsville, MD 20705, USA

T. Sakurai (ed.), *The Hinode Mission*. DOI: 10.1007/978-0-387-88739-5_7

Abstract The EUV Imaging Spectrometer (EIS) on Hinode will observe solar corona and upper transition region emission lines in the wavelength ranges $170-210$ Å and $250-290$ Å. The line centroid positions and profile widths will allow plasma velocities and turbulent or non-thermal line broadenings to be measured. We will derive local plasma temperatures and densities from the line intensities. The spectra will allow accurate determination of differential emission measure and element abundances within a variety of corona and transition region structures. These powerful spectroscopic diagnostics will allow identification and characterization of magnetic reconnection and wave propagation processes in the upper solar atmosphere. We will also directly study the detailed evolution and heating of coronal loops. The EIS instrument incorporates a unique two element, normal incidence design. The optics are coated with optimized multilayer coatings. We have selected highly efficient, backside-illuminated, thinned CCDs. These design features result in an instrument that has significantly greater effective area than previous orbiting EUV spectrographs with typical active region $2-5$ s exposure times in the brightest lines. EIS can scan a field of 6×8.5 arc min with spatial and velocity scales of 1 arc sec and 25 km s^{-1} per pixel. The instrument design, its absolute calibration, and performance are described in detail in this paper. EIS will be used along with the Solar Optical Telescope (SOT) and the X-ray Telescope (XRT) for a wide range of studies of the solar atmosphere.

1. Introduction

The Hinode mission will study the Sun at visible, EUV and X-ray wavelengths. Visible observations will be made with a 0.5 m diffraction-limited telescope — the largest solar optical instrument yet deployed in space. The Solar Optical Telescope (SOT), constructed by NAOJ and Lockheed-Martin, will investigate photospheric dynamics and make vector magnetogram maps at ≈ 0.25 arc sec (175 km) resolution.

X-ray observations will be made with a grazing incidence X-Ray Telescope (XRT) having 2 arc sec spatial resolution. Constructed by Smithsonian Astrophysical Observatory and NAOJ, it images the entire solar atmosphere in the temperature range $1\ \mathrm{MK} < T < 30\ \mathrm{MK}$.

The UK-led EUV Imaging Spectrometer (EIS) will observe the emission lines of highly ionized elements in two carefully chosen wavelength bands so as to measure detailed plasma properties with special emphasis on flow velocities and on non-thermal plasma processes over a wide range of plasma temperatures (0.04 MK, 0.25 MK, $1.0\ \mathrm{MK} < T < 20\ \mathrm{MK}$). This paper outlines the scientific goals of the EIS and discusses the properties, calibration and performance of the instrument in detail within the context of the overall Hinode mission.

R. Moye
Artep Inc., 2922 Excelsior Spring Ct., Ellicott City, MD 21042, USA

H. Hara · T. Watanabe
National Astronomical Observatory of Japan, Mitaka, Tokyo, 181, Japan

K. Matsuzaki · T. Kosugi
Institute of Space and Astronautical Science, Sagamihara, Kanagawa 229, Japan

V. Hansteen · Ø. Wikstol
Institute of Theoretical Astrophysics, University of Oslo, P.O. Box 1029, Blindern, 0315, Oslo, Norway

2. Scientific Aims

The scientific aims of the Hinode mission are focused on three main goals:

- Determine the mechanisms responsible for heating the corona in active regions and the quiet Sun.
- Establish the mechanisms that cause transient phenomena, *e.g.*, flares, CMEs.
- Investigate processes for energy transfer from photosphere to corona.

The instruments have been designed to achieve these goals. Instrument operations and science analysis will concentrate on understanding how changes in the magnetic field impact the solar atmosphere in terms of slow evolutionary behavior, small-scale heating, or through more catastrophic events. In pursuing the mission science goals, recognition of *magnetic reconnection-based* physical processes and their quantitative description will be of considerable importance for understanding the responsive behavior of the solar atmosphere. Another important area is the identification and description of *wave propagation modes* and any related energy dissipation.

The EIS contribution to the mission aims involves the measurement of line intensities, Doppler velocities, line widths, temperatures and densities for the plasma in the Sun's atmosphere. From these measurements, EIS will probe the physical processes that are prevalent on widely different size scales on the Sun. With the availability of suitable multilayer coatings, the design goals of EIS for operation at $\lambda < 300$ Å were to substantially increase the photon throughput and enhance spectral and spatial resolution over previous spectrometers that had operated at these wavelengths. These improvements have led to an instrument that can obtain useful images of an active region (4×8 arc min) at 2 arc sec resolution in around $1-2$ minutes for 12 suitable emission lines. For a flaring active region loop, a 50 Mm section of emitting plasma can be scanned at 2 arc sec resolution in a time of one minute while achieving plasma velocity and line profile width estimates with precisions of ± 5 km s^{-1} and ± 25 km s^{-1} respectively. A selection from the many topics that will be pursued with EIS is indicated below:

Coronal/Photospheric velocity field comparison in active regions: On active region (AR) spatial scales, the visible filter images from the solar optical telescope (SOT) will provide detailed information on photospheric velocities and their time variation. Both vector and line-of-sight magnetograms will also be available. The detailed observation of related intensity, velocity and magnetic configuration changes in the coronal active region plasma has not previously been possible and will be undertaken with EIS observations of loops and other AR magnetic structures.

Coronal AR heating: dynamic phenomena in loops: The understanding of this topic remains elusive. There is evidence for reconnection in loops (*e.g.* Harra, Mandrini, and Matthews, 2004). Much time will be devoted to the detection and characterization of small brightenings by both the EIS and the XRT. EIS in particular will observe any changes in temperature, density and velocity that occur as a result of small events and will obtain evidence of related plasma flows. In addition, the spatially resolved loop temperature and density measurements that EIS will obtain will allow comparison with the output of increasingly sophisticated MHD models (Klimchuk, 2006).

Evolution of trans-equatorial loops: These structures have by definition a significant role for the understanding of large-scale coronal activity. While they appear to participate in large-scale reconnection (Tsuneta, 1996), many of their properties are similar to those of the smaller loops found in active regions (Pevtsov, 2000). Following their recognition in full-Sun XRT images, EIS will study their foot points in an effort to understand their energetics in relation to the underlying magnetic field.

Coronal seismology: waves in AR structures: Waves are observed on all size and time scales on the Sun — from the 5 minute oscillations in the chromosphere to the large-scale shock waves related to flares and coronal mass ejections. They may have a key role in the supply of the energy to the corona and have been demonstrated to exist in coronal structures (*e.g.*, Williams *et al.*, 2002). It is also clear that their detection and mode identification will allow the measurement of important coronal parameters, *e.g.*, magnetic field (Nakariakov and Ofman, 2001). Spectroscopic observations of oscillations in coronal loops have been made in flaring and post flare conditions (Wang *et al.*, 2002, 2003), while on a larger scale spectroscopic observations of EIT waves in the corona have been pioneered by Harra and Sterling (2003). Progress in these important areas requires observations at the better cadence that EIS will provide.

CME onsets and signatures: CMEs almost certainly involve reconnection. Magnetic breakout scenarios (*e.g.*, Antiochos, DeVore, and Klimchuk, 1999), require the removal of overlying magnetic field structures. In the case of eruption of twisted flux ropes (Williams *et al.*, 2005), this removal process permits an eruption that is driven by the energy stored in the twisted field. So far, CMEs have largely been studied with limb observations by coronagraphs. Velocity measurements by EIS will allow the early stages of the field removal to be identified on the disc and the degree of twist in the erupting material to be assessed. Current CME models predict different plasma dynamic signatures. Here again, EIS velocity measurements will have a key role in testing model validities.

Flare produced plasma: source, location and triggering: The production of high temperature plasma in the corona following solar flares continues to be controversial. Bragg spectrometer observations of flare plasma with good spectral resolution by the Yohkoh BCS (Culhane *et al.*, 1991) have had poor spatial resolution. However Warren and Doschek (2005) have reported a hydrodynamic model that involves energy release in successive sub-resolution threads within loops and appears consistent with the plasma velocities observed by the Yohkoh BCS. EIS will image flare lines from *e.g.* Fe XXIV, with good spatial and temporal resolution which, together with XRT context observations, should clarify the plasma production questions.

Flare reconnection: inflow and outflow: While there has been a lot of observational evidence for reconnection in flares (*e.g.*, Masuda *et al.*, 1994; Tsuneta, 1995), there remain inconsistencies in detail. In particular, spectroscopic data are lacking on outflow and inflow velocities. Much evidence for reconnection has been based on imaging alone (*e.g.*, Yokoyama *et al.*, 2001). We require spectral images with high temporal resolution in the corona. EIS is designed to address this difficult problem.

Quiet Sun transient events: network, network boundaries, Coronal Hole boundaries: Evidence has also been found for reconnection in the quiet Sun, around convective cell boundaries (*e.g.*, Innes *et al.*, 1997) and at coronal hole boundaries (*e.g.*, Madjarska, Doyle, and van Driel-Gesztelyi, 2004), resulting in bi-directional jets. Heating has been observed at the cell boundaries (*e.g.*, Harrison, 1997) and even within the cells themselves (Harra, Gallagher, and Phillips, 2000). There is some dispute as to the cause of the bi-directional jets (often termed explosive events) and of the events that are registered through heating or density change (often referred to as blinkers). EIS will enable us to distinguish between these and determine whether they are indeed different phenomena.

The coronal emission lines registered in the two EUV pass bands of the EIS instrument along with the magnetic field data provided by the SOT and the structural context information from the XRT, will allow significant advances to be achieved in the above areas and in many other facets of solar coronal physics.

3. Instrument Overview

Previous spectrometers designed to operate in orbit in the 50 to 500 Å wavelength range have employed grazing incidence optical systems (mirrors and diffraction gratings) since the normal incidence reflectivity at these wavelengths is vanishingly small for the usual optical materials (*e.g.*, SOHO CDS; Harrison *et al.*, 1995). In addition, the microchannel plate array detectors commonly used, although providing good spatial resolution, exhibited quantum efficiencies (QE) $\leq 20\%$ and required hygroscopic coatings, *e.g.*, KBr. Uncoated microchannel plates have substantially lower QE values. The design of the EIS instrument allows normal incidence operation of the optical elements through the use of multilayer coatings applied to both mirror and grating. In addition, the use of thinned back-illuminated CCDs to register the diffracted photons allows QE values to be achieved that are two to three times greater than for microchannel plate systems. A disadvantage stems from the comparatively narrow passband achievable with an individual multilayer. At the time the instrument was designed, the wavelength range obtainable from available multilayers was $80\ \text{Å} < \lambda < 350\ \text{Å}$. However, enhanced knowledge of the coronal emission line spectrum means that these limitations can be tolerated in the interest of achieving high throughput.

The optical design and layout of the instrument are shown in Figure 1. The spectrometer has a large effective area in two EUV spectral bands through the use of Mo/Si multilayer coatings optimized for high reflectivity in the ranges $170-210$ Å and $250-290$ Å. Solar radiation enters through a thin 1500 Å Al filter which stops the transmission of visible radiation. Photons are focused by the primary mirror onto a slit and are then incident on a toroidal concave grating. Two differently optimized Mo/Si multilayer coatings are applied to matching halves of both mirror and grating.

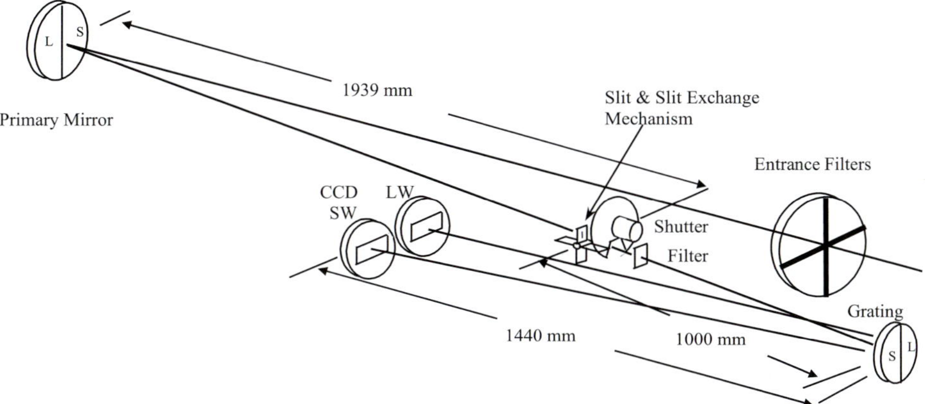

Figure 1 Optical layout of the spectrometer. Components are labeled and dimensions are given in mm. S/SW and L/LW refer to short and long wavelength bands.

Table 1 EIS performance parameters.

Wavelength bands	170–210 Å and 250–290 Å
Peak effective areas[a]	0.30 cm^2 and 0.11 cm^2
Primary mirror	15 cm diameter; two Mo/Si multilayer coatings
Grating	Toroidal/laminar, 4200 lines/mm, two Mo/Si multilayers
CCD cameras	Two back-thinned e2v CCDs, 2048 × 1024 × 13.5 μm pixels
Plate scales[a]	13.53 μm/arc sec (at CCD); 9.40 μm/arc sec (at slit)
Spatial resolution (pixel)	2 arc sec (1 arc sec)
Field of view	6 arc min ×8.5 arc min, offset center: ±15 arc min E-W
Raster	1 arc sec in 0.7s[b] (Minimum step size: 0.123 arc sec)
Slit/slot widths[a]	1, 2, 40 and 266 arc sec
Spectral resolution[a]	47 mÅ (FWHM) at 185 Å; 1 pixel = 22 mÅ or approx. 25 km s^{-1} pixel^{-1}
Temperature coverage	Log $T = 4.7, 5.6^c, 5.8^c, 5.9^c, 6.0–7.3$ K
CCD frame read time	0.8 s
Line observations	Simultaneous observation of up to 25 lines

[a]Measured values.

[b]Raster steps occur during CCD readout sequences.

[c]Quiet Sun lines of Fe VIII, Si VII, Si VIII; count rates ≈0.5–1.0 counts s^{-1} pixel^{-1}.

Diffracted radiation is registered by a pair of thinned back-illuminated CCDs. Exposure times are controlled by a rotating shutter while a slit exchange mechanism can allow selection of four possible apertures — two spectral slits and two spectral imaging slots. A second thin Al filter is mounted behind the slit/slot mechanism to provide redundancy. The larger entrance filter is housed in an evacuated enclosure that will give protection from acoustic stress and debris during the launch. Raster scanning capability is provided by a piezoelectric drive system which rotates the primary mirror. The raster scan range of 6 arc min in the dispersion direction and the useable slit height of 8.5 arc min set the overall instrument field of view. In addition, there is a coarse mechanism that can offset the mirror by ±15 arc min from the spacecraft pointing axis in an E-W direction. The grating has a focusing mechanism for on-orbit adjustment. The instrument properties are summarized in Table 1.

All of the components shown in Figure 1 are mounted in a composite structure which, because of its low Coefficient of Thermal Expansion (CTE), acts as a stable bench for the optical components. A cooling radiator on the outside of the structure maintains the CCDs at $<-50°$C at expected operating conditions. With the front filter enclosure doors closed, the interior of the spectrometer can sustain a slight positive pressure when under dry nitrogen purge. The latter is maintained throughout the period when the instrument is on the ground and is only to be removed about two hours before launch when the spacecraft is inside the fairing of the third stage.

An electronic overview of the EIS instrument is given in Figure 2. The primary digital and power interfaces between EIS and the spacecraft are handled by the Instrument Control Unit (ICU). There are three digital interfaces, command, housekeeping (status) and science (mission data). All three have serial differential format. The ICU hosts compiled C software running on a TEMIC 21020 Digital Signal Processor (DSP) that interprets commands and controls the instrument engineering and science operations. The software gathers housekeeping and returns it to the spacecraft, monitoring the health of selected parameters and

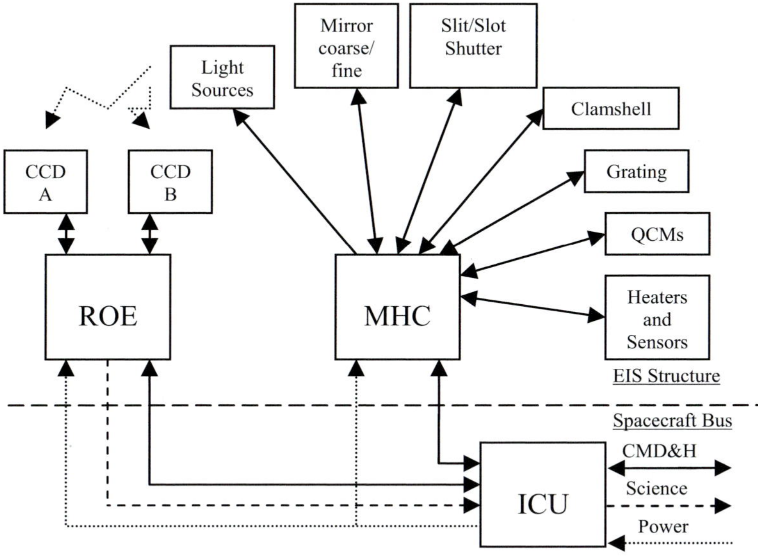

Figure 2 EIS electronic diagram showing the instrument subsystem interconnections.

putting the instrument into a safe mode should any exceed defined limits. The software also packetizes science data and sends them to the spacecraft via the science interface.

The power interface is a 28 V regulated bus. The ICU contains a DC/DC converter which supplies isolated secondaries for the ICU and the camera Read Out Electronics (ROE). No switching power conversion is done in the ROE to minimize local noise generation. The 28 V power for the Mechanism and Heater Control (MHC) unit is controlled by switches in the ICU. The MHC houses its own power converters.

Digital communication for commanding and housekeeping between units within the instrument is facilitated by separate RS422 interfaces between the ICU and ROE and between ICU and MHC. For the science data there is a dedicated 32 Mbps link from the ROE to the ICU. The ROE interfaces to the CCDs and the MHC interfaces to the EIS structure mechanisms, quartz crystal microbalances (QCMs), CCD light sources, heaters and temperature sensors. Heaters to replace the heat dissipated by the ROE and MHC when they are off (make-up heaters) along with CCD bake-out heaters and temperature sensors for decontamination of the CCDs are operated by the ICU. In addition to the above ICU interfaces, there are direct connections from the spacecraft to the spectrometer structure for temperature sensors and survival heater circuits.

The EIS grounding scheme is chosen to give the best ROE noise performance. There is a single point ground for the EIS ICU and ROE secondary power which is at the ROE. There is also a single point ground for the MHC power systems via a link at the spectrometer Connector Panel (CNP). The ICU/ROE and MHC grounds are joined together at the CNP which is then connected to the spacecraft ground. The composite structure resistance is not low enough to guarantee a Faraday cage effect so the spectrometer harness is screened. The electrically conducting Multi-Layer Insulation (MLI) external to the structure also helps to screen the internal electronics.

The CCD camera electronics and the MHC unit are located within the spectrometer section. The independent EIS ICU and the spacecraft Mission Data Processor (MDP) are located in the spacecraft service or bus section at $\approx$2.5 m from the spectrometer. Observation tables, loaded in the EIS ICU from the MDP, organize the readout of data from the CCD cameras with a maximum of 25 user-selected spectral windows being allowed by the software. The ICU also generates commands for operating the scanning, shutter, and slit interchange mechanisms to execute appropriate sets of observing sequences or studies. The spacecraft mass memory has a total capacity of 7 Gbits for the instruments and the nominal EIS share of this is 15%. Thus instrument data throughput is set by the number of ground station contacts per day. Provision of access to the Norwegian Svalbard ground station by ESA and the Norwegian Space Agency will allow at least 15 ground station contacts each day in addition to the four daily contacts with the JAXA station at the Uchinoura Space Centre (USC). Thus EIS, using lossless data compression, will be able to operate at a data rate of $\approx$100 kb/s.

Following the SOHO CDS instrument, EIS will provide the next steps in EUV spectral imaging of the solar corona and upper transition region. It will have approximately a factor of ten enhancement in effective area due to the use of multilayer coated optics and back-illuminated CCDs. Spectral resolution is also improved by a factor of ten in the wavelength ranges being observed. While at 2 arc sec, the spatial resolution is a factor two to three better than that of CDS.

4. Optical Design and Instrument Components

An optical schematic of EIS which gives the locations of the components is shown in Figure 1 and has been briefly discussed in the previous section. A detailed account of the instrument's optics and mechanisms is given by Korendyke *et al.* (2006). The telescope primary mirror images EUV radiation from the Sun onto the spectrograph slit. Light passing through the slit is dispersed and stigmatically re-imaged by the toroidal grating onto two 1024×2048 pixel CCD detectors, each with 2048 pixels in the dispersion direction. In flight, the mirror can be rotated in $\approx$0.125 arc sec steps about the Y axis (solar N-S) to sample different solar structures with the slit. High-resolution spectroheliograms (raster images) are formed by steadily moving the solar image in fine increments on the spectrograph slit and taking repeated exposures. An interchange mechanism allows selection among two slits (1 and 2 arc sec width) and two slots (40 and 266 arc sec width). The slot observations of the solar disk obtain images of large areas in bright solar emission lines with a single exposure. For the 40 arc sec slot, spectrally pure images are available for several strong lines in each passband. The slot images exhibit a modest spatial blur along the dispersion direction.

Both mirror and grating operate at near normal incidence. To broaden the spectral range, the multilayer-coated optical elements were divided into two D-shaped sectors; each sector was coated with a multilayer tuned to produce high reflectivity in its wavelength band. These coatings achieved peak reflectivities of 32% and 23% in the 170 – 210 Å and 250 – 290 Å bands, respectively. Optimum response is achieved for each band by careful selection of thickness for the individual Si and Mo layers (Seely *et al.*, 2004). The telescope mirror is a superpolished off-axis parabola with a focal length of 1939 mm, a measured figure accuracy $<\lambda/47$ rms at 6328 Å and a microroughness of <4 Å rms. The 160 mm diameter mirror was fabricated from Zerodur by Tinsley Laboratories, Inc. and has a usable diameter of 150 mm. The flight grating is specified to be a toroid (Beutler, 1945; Haber, 1950) with radii of 1182.98 mm in the dispersion direction and 1178.28 mm in the perpendicular direction with a figure slope error <0.5 arc sec RMS. The grating microroughness is <2.5 Å RMS. The gratings were fabricated by Carl Zeiss Laser Optics GmbH from 100 mm diameter fused silica blanks and have a usable area of 90 mm diameter.

4.1. Entrance Filter and Housing

In the 1970s, NRL in collaboration with G. Steele of Luxel Corp. developed large format thin aluminum filters for the EUV instruments on Skylab (Schumacher and Hunter, 1977). High purity vapor deposited aluminum (VDA) foils between 1000 Å and 1500 Å thick pass EUV wavelengths between 170 Å and 650 Å while blocking visible, infrared and near UV wavelengths (Powell, 1992). We have implemented a set of four Al filters of this type for the EIS front aperture and a second smaller Al filter behind the spectrometer slit to reject out of band light that might leak through pinholes. For the launch, the thin foils of the entrance filter must be protected from severe vibration, the acoustic and debris environment, and from oxidation and contamination. A special chamber was built for launching the filters under vacuum (Figure 3). It is a short cylinder having front and rear doors hinged to open in the vacuum of space much like a clam and is thus named the Clamshell (CLM) assembly. In orbit, the doors are opened by high output paraffin (HOP) thermal actuators.

The CLM and filter arrangement evolved from similar units used on the TRACE mission (P. Cheimets, 1999, private communication), with the front aperture divided into four quadrants, each with its own filter (Figure 4). The thin Al filters are supported on a nickel mesh of 40 μm wires on 390 μm centers. Each mesh has an open area equal to 80% of the total. Meshes with attached foils were glued to Al frames for installation. Each frame has a blackened tongue and groove air passage around its margin to allow air pressure to equalize on the two sides of the filter.

The EUV transmission of each filter was measured with synchrotron radiation at the X24C beamline of the National Synchrotron Light Source (NSLS) at the Brookhaven National Laboratory. The synchrotron beam passed through a monochromator and sampled an ≈2 mm diameter spot on the filter. Each filter was measured at a grid of points on 15 mm spacing. Figure 5 shows a summary plot of the transmission of a typical filter, in this case

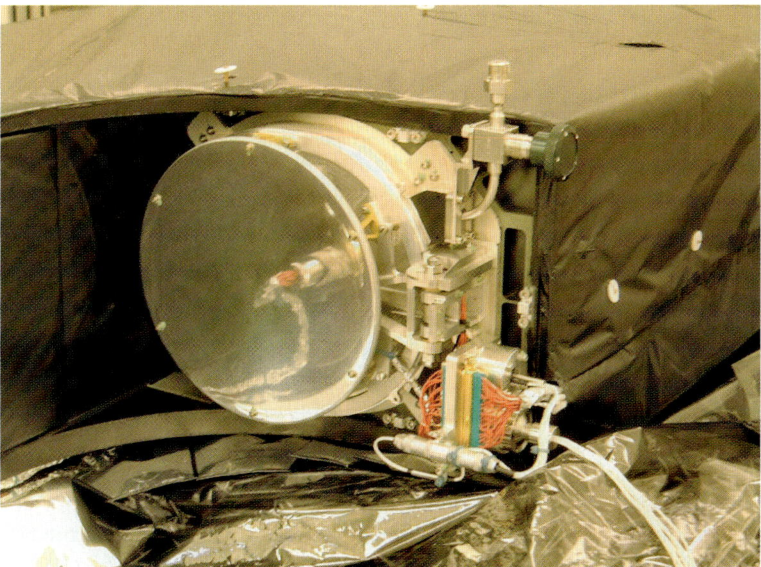

Figure 3 EIS Clamshell assembly mounted on instrument. The evacuation port is at the top and a thin polished metal sun shield disk on the front reflects incoming solar radiation. The instrument itself is covered with black thermal blanketing material.

Figure 4 Engineering model of EIS front filter array. Four frames are in place. This view of the array is before the aluminum foils were mounted.

Figure 5 Average transmission for a 1500 Å Al filter. The data are from 24 scans at points on a 15 mm grid arranged over the filter surface. The error bars are $\pm 1\sigma$. These data are for the mounted filters and include the $\approx 15\%$ loss due to the mesh. The step at 170 Å is due to the Al absorption L-edge.

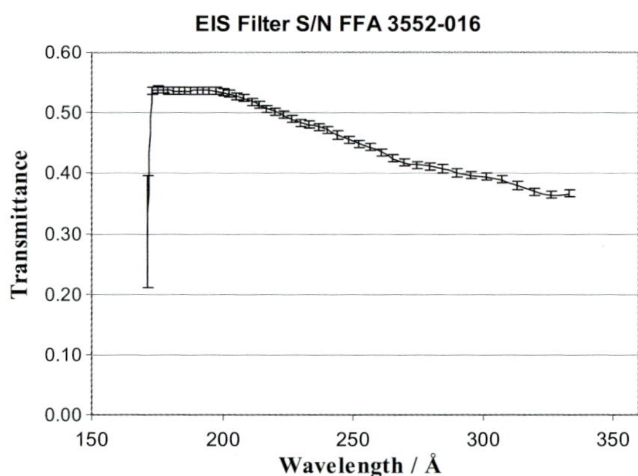

1500 Å thick. Another filter property measured was the visible light rejection. The visible transmission fraction was $\leq 8.3 \times 10^{-7}$ for all filters.

4.2. Primary Mirror and Scanning Mechanisms

The articulated primary mirror produces a high quality solar image at the spectrograph entrance slit. The mirror mechanism provides two different types of articulation for moving the solar image perpendicular to the slit. A piezoelectric transducer (PZT) actuator and flex-pivot arrangement provides a tilt motion about a N-S line through the mirror center. The maximum mirror tilt is 300 arc sec and the image motion at the slit is therefore 600 arc sec. The measured reproducibility of the mirror movements was better than 2 arc sec over about 30 minutes time in a laboratory environment. Nominal rastering operations will be conducted by tilting the mirror in fine increments with the PZT. This produces an E-W motion of the solar image through a small range. Linear mirror motion is provided by a ball screw

Table 2 Predicted EIS mirror imaging performance.

Contributing error	Applicable tolerance	Spot diameter [μm]	Comments
Optical aberrations	<150 μm along axis	<13.5	Modeled geometric and diffraction effects
PZT electrical noise		<7.75	Measured
Focus error budget		<11.1	Sensitivity of focus test
Total		19.12	Root sum square

Note: Equivalent spot diameters are for CCD focal plane. Diameters twice rms radii.

and linear bearing arrangement. The step size is equivalent to 0.30 arc sec on the solar surface with a total range greater than ± 2750 steps or ± 825 arc sec. Linear mirror travel is perpendicular to the chief ray from the mirror center to the slit center. In orbit, the linear motion will be used to re-point the instrument field of view. It also moves the solar image in the E-W direction on the slit, but the range is larger than that of the PZT tilt. The linear motion of this mechanism may disturb the fine pointing of the Solar B spacecraft and hence its operation is restricted. However, the EIS instrument can use this mechanism to observe emerging activity at the limbs on occasions when the SOT may be studying mature active regions at disc center.

The predicted image quality of the mirror is shown in Table 2. The ray trace spot diameters degrade slowly as a function of field angle and mirror position. For a nominal raster using the PZT with a field of view of ± 250 arc sec along the slit and a total raster width of ± 250 arc sec, the rms spot diameter is 0.8 arc sec at the corners. A linear excursion of the primary mirror of 900 arc sec results in a ray trace spot diameter of 1.9 arc sec at the center of the slit. During normal operations, the maximum expected repointing adjustment is <2 arc min with spot diameters <0.3 arc sec. Overall, the expected ray trace/geometric spot diameter is <0.7 arc sec within a field of 250 arc sec radius. The EIS primary mirror with a $\lambda/47$ rms reflected wavefront error operated at EUV wavelengths will achieve ≈ 0.6 arc sec diameter diffraction limited performance. A more complete discussion of diffraction theory of aberrations is given in Born and Wolf (1964). With equivalent diffraction spot diameter of ≈ 0.6 arc seconds and expected ray trace spot diameters of <0.7 arc sec, we expect to achieve <1 arc sec spot diameters from the mirror optic. The rms microroughness (<0.3 nm) is sufficient to obtain high quality EUV imaging. Typical large telescope microroughness in the visible is 2 nm rms, which is comparable to the scaled situation in EIS. The wings and overall shape of the point spread function at the slit should be similar to that of present orbiting telescopes (TRACE and EIT/SOHO) with prefilters. The telescope imaging was verified to be <2 arc sec in the flight configuration. We illuminated the front aperture with visible light using a high quality collimator and examined the resulting image at the slit plane with a microscope arrangement. We were able to clearly discern <2 arc sec resolution on the projected and subsequently re-imaged 1951 USAF resolution target. A similar optical setup was used to focus the telescope.

The components of the mirror mechanism are shown in Figure 6. The central hub is bonded to a bracket with flexible epoxy. The bracket is mounted on two flexible pivots and attached to the linear moving stage of the mechanism. The PZT actuator provides the rotary (mirror tilt) motion. Strain gauges in the PZT sense the extension and remove hysteresis. Open and closed loop PZT performance is shown in Figure 7. The linear moving stage

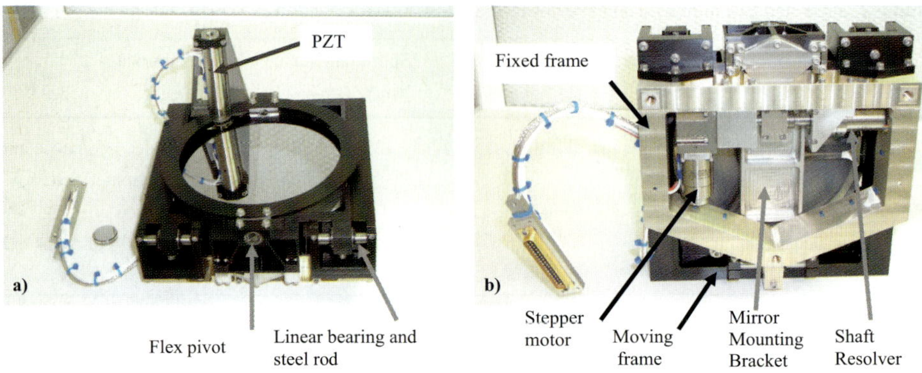

Figure 6   The assembled flight mirror subsystem (a) front view and (b) rear view.

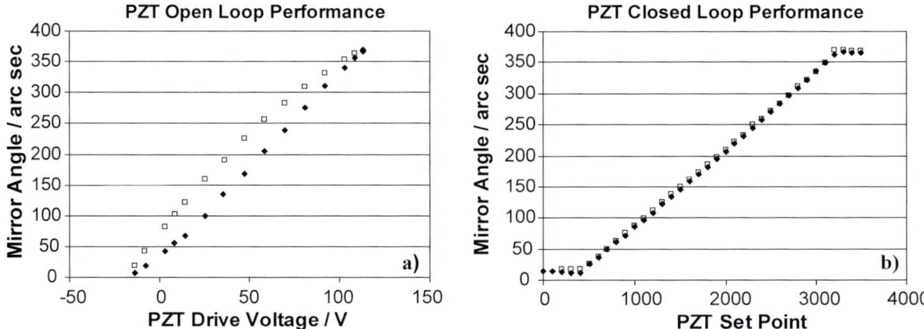

Figure 7   Open (a) and closed loop (b) PZT performance. Rotation angle plotted as a function of control loop set point and as a function of PZT voltage. Solid diamonds are for increasing PZT voltage and open squares are for decreasing PZT voltage. The useful PZT set point range is between 600 and 3200.

is attached to a fixed frame with three linear, self-aligning bearings mounted on hardened steel shafts. A linear actuator, attached to the fixed frame, consists of a 30 degree stepper motor with a gearbox and ball screw. It drives a ball nut attached to the moving frame. The residual displacement errors are periodic in nature with maximum amplitude of ±15 µm or 1.6 arc sec on the solar surface.

4.3. Shutter and Slit Exchange Mechanisms

The EIS instrument incorporates a slit/slot interchange mechanism shown in Figure 8a. This mechanism enables selection of one of four instrument slits/slots to support various observation programs. Each slit and its aluminum frame are bolted to a paddle wheel. Tight fabrication tolerances on the frame and paddle wheel control the placement of the slits. The paddle wheel (Figure 8b) is directly attached to the output shaft of a rotary actuator whose axis of rotation is perpendicular to the chief optical ray. The rotary actuator reproducibly places each of the spectrometer slits in the telescope focal plane. A geared rotary resolver is attached to the output shaft of the stepper motor and has an accuracy of better than 15 arc min, sufficient to discriminate between individual motor steps. The instrument slits and slots were fabricated by etching from a silicon substrate. Precise metrology was carried out to determine the slit width, results of which are shown in Table 3.

Figure 8 (a) Slit/slot mechanism assembly. The four slits are on a paddle wheel and are exchanged by 90 degree rotations of the wheel. The shutter blade and motor are also attached to this assembly. (b) Paddle wheel and slit frame before blackening.

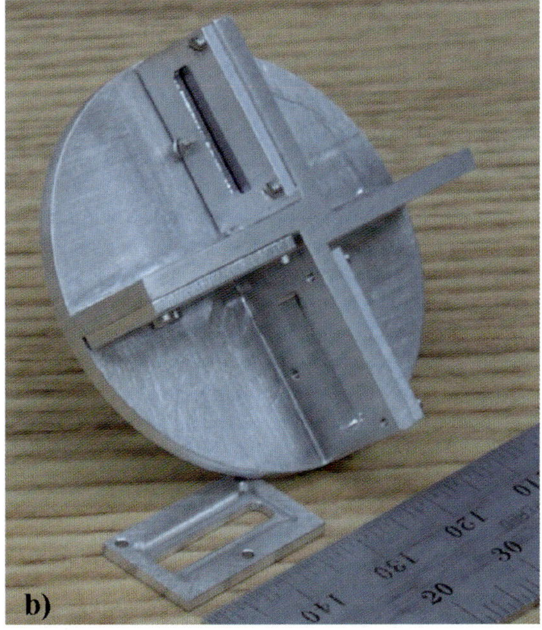

Table 3 EIS slit width summary.

Nominal width [arc sec]	Serial number	Measured width [μm]	Measured width [arc sec]
1	101C	9.5	1.01
2	102A	19.0	2.02
40	103A	384	40.9
266	104D	2506	266.6

Note: Measured accuracies are ±0.5 μm for the 1 and 2 arc sec slits and ±2 μm for the 40 and 266 arc sec slots.

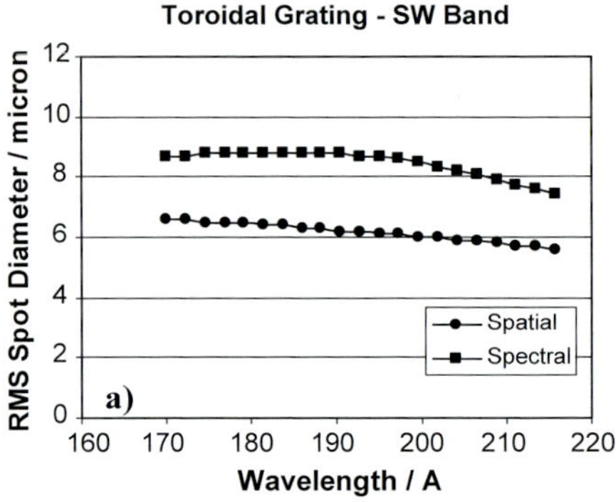

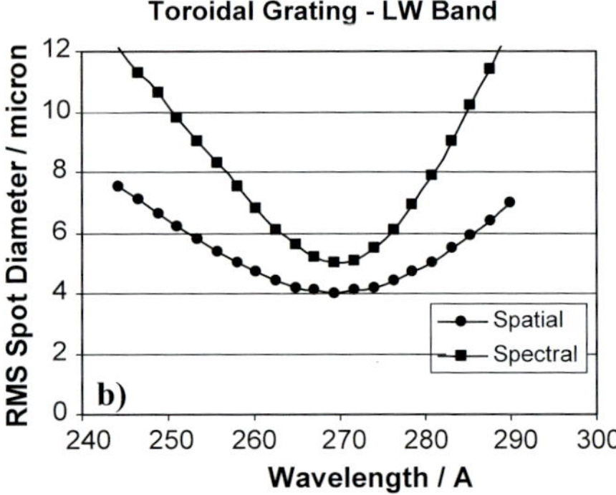

Figure 9 Grating spot size diameter for a point at the center of the slit as $f(\lambda)$ for (a) short wavelength band and (b) long wavelength band.

As shown in Figure 8a, the shutter assembly is combined with the slit interchange assembly. During laboratory testing, the shutter and associated flight drive electronics successfully obtained a series of 100 ms exposures with <5% photometric error. The shutter motor is a brushless DC unit available from Kollmorgen. The necessary commutation is provided by an optical encoder assembly mounted to the shaft of the motor. The motor was specifically designed to provide the high speed and high torque for precise control of rotating assemblies required by this application.

4.4. Concave Grating

The grating mount geometry was optimized using a ray tracing program. The grating ruling frequency was chosen to be 4200 lines/mm, being the limit of the holographic recording equipment at the time of fabrication. The general constraints included an overall instrument length of ≈3 m, a required plate scale of 1 arc sec/pixel, and 13.5 μm CCD pixels. Within these limitations, a ray tracing code optimized the grating parameters to provide the smallest averaged RMS spot size over the spectral bands 170–210 Å and 250–290 Å and over a spatial field of view of ±250 arc sec. The optimization calculation used separate D-shaped portions of the grating for each wavelength range but was constrained to use an identically figured and singly ruled grating for the entire EIS range. This optimization step resulted in a different optical prescription and smaller spot sizes when compared to gratings optimized using the full grating aperture for both wavelength ranges. Representative plots of the rms spot diameters against wavelength and field angle along the slit are given in Figures 9 and 10. Performance as characterized in the laboratory closely approaches these theoretical values. The predicted grating imaging performance is included in Table 2.

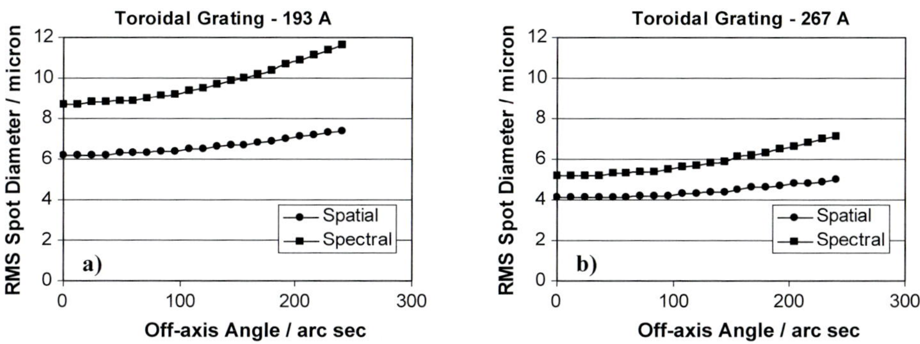

Figure 10 Grating spot size diameter along the slit at the central wavelength of each detector. (a) for the short wave band and (b) for the long wave band. Position 0 is slit center and the off-axis angle is measured North or South along the slit.

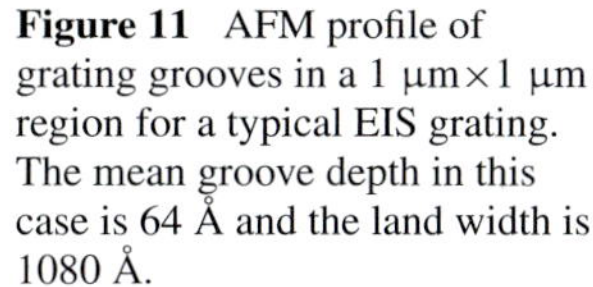

Figure 11 AFM profile of grating grooves in a 1 μm×1 μm region for a typical EIS grating. The mean groove depth in this case is 64 Å and the land width is 1080 Å.

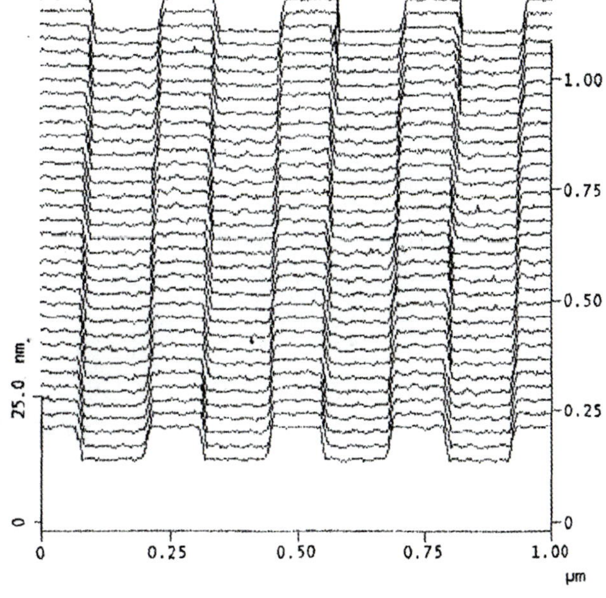

The grating pattern was holographically recorded and subsequently ion etched into the fused silica blank. A single uniform laminar line pattern, with a frequency of 4200 lines/mm, groove depth of 60 Å, and land to groove ratio of 0.85:1 was chosen to provide nearly equal diffraction efficiencies in the two EIS spectral bands. This grating configuration illuminated at near normal incidence should maximize the diffraction efficiency in the first order, while minimizing the efficiencies in other orders (Seely *et al.*, 2004; Kowalski *et al.*, 1999). Figure 11 shows the results of an Atomic Force Microscope (AFM) measurement of the groove profile of an EIS laminar grating. The multi-layer coated grating efficiencies for the zero and first orders are shown in Figure 12. The peak first order efficiencies are 8.0 % at 196 Å in the short wavelength band and 7.9 % at 271 Å in the long-wavelength band.

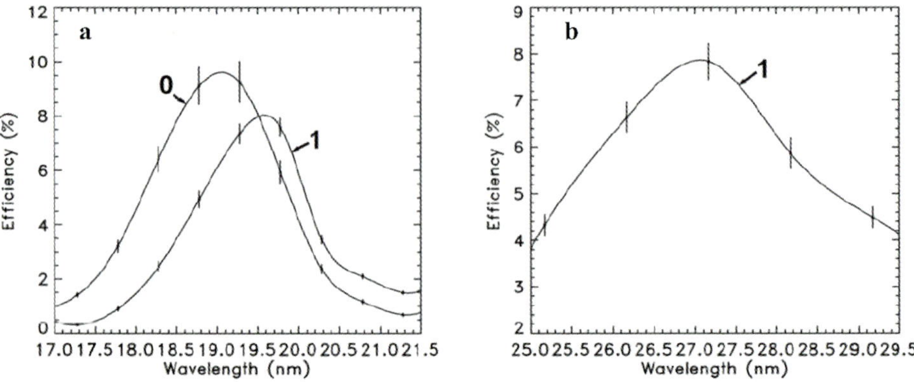

Figure 12 Grating efficiency in the short (a) and long (b) wavelength bands. Grating order number (*m*) is indicated. For *m* greater than 2–5, the measured grating efficiency is <1%.

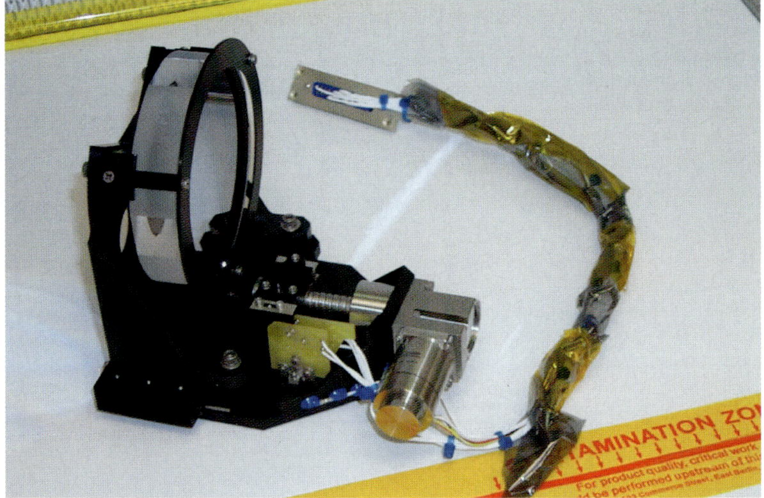

Figure 13 Grating FL-7 in its flight mounting. Two of the four alignment facets on the margin of the grating blank are visible. The focus mechanism includes motor, gearhead, and ball screw. Two small circuit boards hold LEDs and photodiodes for the limit sensors.

The EIS grating mounted in its focusing mechanism is shown in Figure 13. The rear surface of the grating is bonded to a mounting bracket at the grating center with the same flexible epoxy used for bonding the mirror. The grating and bracket subassembly are attached to a moving stage mounted to the mechanism base plate with a pair of crossed roller bearing slides. The moving stage is driven by a geared stepper motor and ball screw combination. Each motor step results in a 2.8 μm displacement of the grating. The mechanism is aligned to drive the translation stage along the slit-grating axis.

4.5. Dual CCD Camera and Readout Electronics

The EIS camera consists of the Read-Out Electronics (ROE) unit and two CCDs, on a Focal Plane Assembly (FPA), at the focus of the spectrometer. Two CCDs are required to cover both the short (170 to 210 Å) and long (250 to 290 Å) wavelength ranges. Radiation from the solar region of interest is focused and dispersed by the EIS optics (Figure 1) into these two wavelength ranges and imaged by the two CCDs.

The EIS CCDs are type CCD 42-20, made by e2v Technologies. They are back-illuminated and thinned to maximize the quantum efficiency. There are 2048 imaging pixels in the dispersion direction and 1024 pixels in the spatial direction. They are three-phase devices which operate in full-frame mode with readout ports at both ends of the readout registers. Pixel size is 13.5 μm × 13.5 μm, which is equivalent to 1 arc sec in the spatial dimension and 0.0223 Å in the spectral dimension. The CCDs are made for Advanced Inverted Mode Operation (AIMO) which allows low dark current levels to be achieved without excessive cooling. For EIS the requirement is −40 °C and an operational temperature of < −45 °C is predicted from modeling for all operational conditions. A single dump gate allows the CCD to be rapidly flushed while clocking the CCD vertically. The relatively high electrical capacitance to substrate of the vertical electrodes places a limitation on the rate that the vertical electrodes may be clocked (8 μs/row). However, it is the horizontal readout rate (2 μs/pixel) which dominates the overall readout duration rather than the vertical clock rate. The CCD type 42-20 has 50 overscan pixels at each end of its readout register. On-chip horizontal binning is achieved with a summing well electrode at each end.

The CCD mounting is shown in Figure 14a. Devices are mounted on INVAR plates which have a low CTE of $(1.3 \times 10^{-6}\,°C^{-1})$ which is reasonably matched to that of the silicon imaging surface $(2.6 \times 10^{-6}\,°C^{-1})$. The INVAR plates are mounted on titanium barrels which are attached to the main camera bracket. A Vespel spacer is located between the titanium barrel and the structure to thermally isolate the CCDs. The INVAR plates also support the associated thermal sensor and survival and bake-out heaters. The devices are connected to the ROE by flexible cables and micro-D connectors. The CCD parameters are summarized in Table 4.

For the EIS camera, having ports at each end of the readout registers not only speeds up the overall readout time but also allows for some redundancy so that if one port fails, the ROE can be programmed to read out pixel data from the remaining functional port. Thus

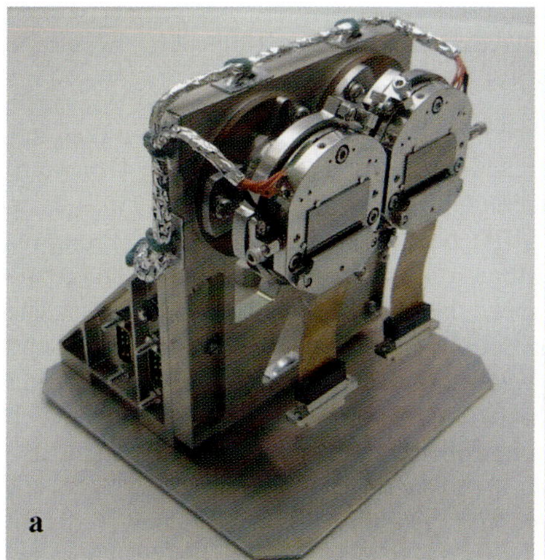

Figure 14 The EIS dual CCD system with (a) the CCDs mounted showing their flexible electrical connections, and (b) the flight readout electronics unit. Cooling strap connections to the CCDs are shown attached to the particle/radiation shield on top of the ROE box.

Table 4 Summary of CCD parameters.

Wavelength range	CCD A 250 – 290 Å	CCD B 170 – 210 Å
Device type	CCD 42-20	
Array size	2048 × 1024	
Pixel size	13.5 μm × 13.5 μm	
Readout rate	2 μs/pixel	
Full well capacity	90k electrons	
Charge transfer efficiency	0.999996	
Quantum efficiency[a]	39 ± 4%	44 ± 4%

[a]See Section 7.2. Full calibration results suggest that ≈60% may be more appropriate.

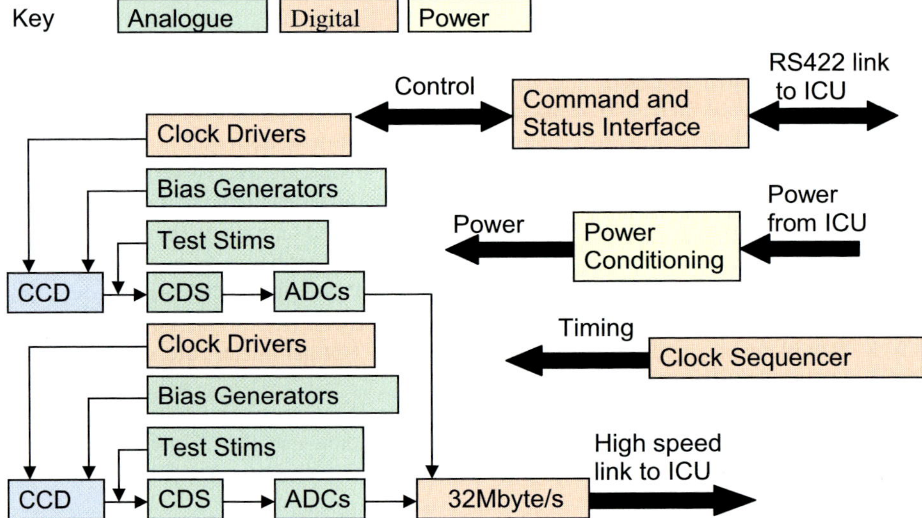

Figure 15 Block schematic diagram of the EIS dual-CCD camera and its readout electronics.

the ROE requires four analog signal chains in order for all four ports to be read out simultaneously. The organization of the camera and readout system is shown as a block diagram in Figure 15. A Correlated Double Sampling technique reduces the CCD reset noise to an acceptable level. Each signal chain produces 14-bit parallel data which a Field Programmable Gate Array (FPGA) combines into a 32 Mbps serial stream. This is transmitted to the EIS Instrument Control Unit (ICU) via a high-speed data link in the telescope harness. The Clock Sequence Generator (CSG) or sequencer is common to both CCDs for simplicity and elimination of cross-talk. There is a local test CSG in the analog electronics for pre-flight testing and noise performance evaluation, and a main CSG in the digital electronics which is programmable from the ICU. The main CSG controls the clocks for the CCD and can also generate stimulation test patterns.

The ROE hardware consists of a motherboard and three daughter boards: power, digital and analog. There are two types of serial link with the ICU. A low-speed (9.6 kbps) bi-directional asynchronous command and status link sends commands to the camera. These include master reset, integration time, CCD window origin and size, and the ports used for read-out. The link receives digitized analog housekeeping and camera status information.

A high-speed link (32 Mbps) using Low Voltage Differential Signaling technology passes the CCD image data to the ICU. The 14-bit data from the four CCD signal chains plus a 2-bit CCD port ID header for each is concatenated into a contiguous 64-bit serial word.

The full well capacity of the CCDs is approximately 90k electrons. With 14-bit Analog to Digital Converters (ADCs), the amplifier gain was set to 6.6 ± 0.03 electrons per Data Number (DN). The full well capacity of the CCD in the long-wavelength range corresponds to ≈ 7500 photons. Due to the photoelectric effect, an incident photon will be converted into a variable number of photoelectrons. At 170 Å, each photon will generate about 20 photo-electrons, while at 290 Å, one photon will generate about 12 photoelectrons. The minimum signal detectable by the CCDs should correspond to one photon which in turn corresponds to 12 photoelectrons at the long-wavelength limit. Thus, the minimum "shot" noise on the detected signal should also correspond to 12 photoelectrons or one photon. To maximize the signal-to-noise ratio, the quantization and read noise values should be below the signal shot noise. Since these terms add in quadrature, there is little advantage in having quantization or read noise values which are substantially below the photon shot noise.

The thermal noise generated at the expected nominal CCD operating temperature of $-50\,^{\circ}$C will be minimal (≈ 0.005 electrons pixel^{-1} s^{-1}) except for very long integration times, and will even then still be significantly below the signal shot noise. An amplifier gain of ≈ 6.5 electrons/DN means that the quantization noise is well below the signal shot noise. A readout noise of $\approx 50\%$ of the signal shot noise would correspond to about 6 electrons rms, suggesting a readout time of around 2 μs/pixel.

Subsets of the CCD frames can be selected in hardware for readout. Definable regions on the CCDs are in the form of rectangles selectable up to the full CCD width. Both CCDs must have identical windows which can be up to 1024 pixels wide for two readout nodes or up to 2048 pixels if only one readout node is available. Thus either two or four identical hardware windows can be selected. In the four-window case, pairs of windows must be located symmetrically about column number 1024 on each CCD. The available height is 512 rows. Charge from outside these windows will not be read and can be quickly dumped. A common Clock Sequence Generator leads to the use of identical hardware windows on both CCDs. However, this gives a major benefit of minimizing cross talk on the digitized signal as pixels are clocked out at the same time. Flexible software windows, up to a maximum of 25 in total, can be set to further reduce the pixel data for transmission to the ground. The configuration of these software windows can be independent of CCD and at any location within the frames. Thus particular EUV spectral regions of interest may be selected.

5. Mechanical and Thermal Design

5.1. Mechanical Design

The mechanical design requirements were (a) total structure mass of less than 23 kg set by the capacity of the M-V launcher; (b) high stiffness with a first resonance mode frequency >60 Hz; (c) dimensional stability to maintain spectral and spatial resolution over a broad temperature range and to minimize motion of the spectra on the detectors; and (d) structural three-year condensable molecular fluence of 2.7×10^{-6} g cm^{-2}. This strict fluence requirement restricted the selection of materials within the optics cavity, mandated extensive vacuum conditioning of the EIS structure, and required assembly in a carefully controlled clean environment. The assembled instrument was double bagged and purged during instrument and spacecraft level integration and test.

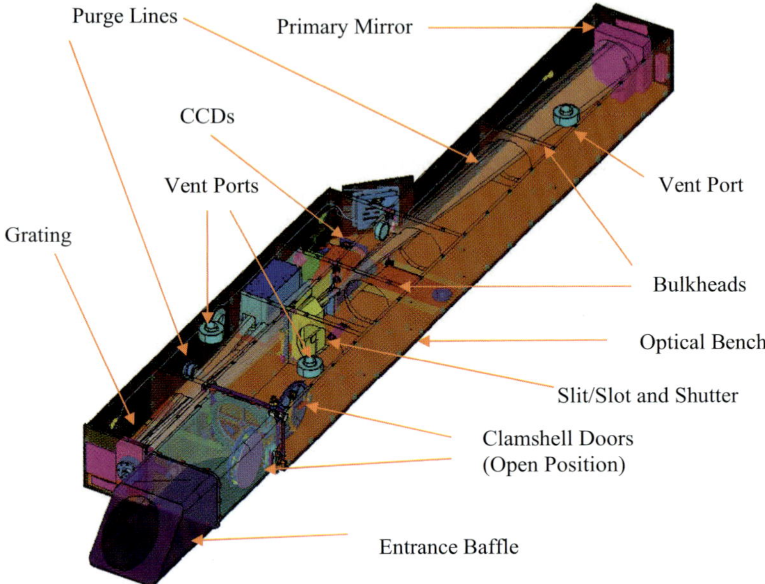

Figure 16 Diagram of the EIS spectrometer structure showing the locations of the principal elements.

The structure chosen used honeycomb sandwich panels with aluminum core material and face sheets made from Carbon Fiber Reinforced Plastic (CFRP). The selected fiber resin combination is M55 with RS3 cyanate ester resin. M55/RS3 has been used extensively in the aerospace industry and shows impressive stiffness to mass ratios with excellent dimensional stability and acceptably low outgassing properties.

The layout of the instrument is shown in Figure 16. EIS measures 3.54 m from the tip of the entrance baffle to the rear bulkhead carrying the mirror. The instrument base panel consists of a 20-mm-thick Al core with nominal 1-mm-thick face sheets and 1-mm reinforcement patches near the instrument interface points. The EIS is mounted on a semi-kinematic titanium suspension between the instrument interface points (titanium inserts) and the central cylinder of the payload module. The width at the widest point of the instrument is 0.55 m and the height is 0.25 m.

The effective CTE of the optical bench was measured using coupon samples and is less than 0.4 ppm/°C. The allowed thermal gradient variation in orbit with this CTE is 10 °C. The bulkhead skins consist of balanced quasi-isotropic layers of 0/+60/-60 deg orientation. The instrument sidewalls and bulkheads consist of 10-mm-thick core with 0.6-mm-thick face sheets. The mass of the bare structure totals just below 23 kg which is 40% of the overall instrument mass (including all optical units, electronics boxes, wiring harness and thermal hardware).

The assembly of the instrument's panels was done in stages to ensure proper cleanliness and stress relief before final integration. All panels including the base panel or optical bench were thoroughly cleaned as parts and then "dry" assembled leaving small clearances between the internal bolted faces and edges. All edges were capped with CFRP U-shaped strips to close the vented aluminum honeycomb core. The face sheets facing outwards were all perforated with a rectangular pattern of small holes. This is to ensure that all gas trapped inside the sandwich was vented directly to the outside of the instrument. The Al honeycomb material was also perforated to permit venting of the individual cells.

The dry assembly was thermally cycled in a vacuum chamber to stress-relieve the structure. Following this step, all strength- and stiffness-critical bulkheads were bonded together with strips of L-shaped CFRP. The edges that were prone to rubbing due to mechanical loading were coated with adhesive to seal the fiber particles on the machined edges. The structure was then returned to the bake-out chamber for a six-week vacuum bake-out with constant QCM monitoring, to clear all outgassing species. After verification and confirmation that the structure met the outgassing fluence requirements, the optical components and electronics boxes were mounted. These were all baked out separately prior to final assembly.

Following final integration and alignment verification the instrument was kept purged during all stages, except during mechanical acceptance, thermal balance and thermal vacuum testing. The purge lines are integrated in the structure and inject ultra-clean nitrogen near all optical components (mirror, grating and CCDs). Vent ports located in the lid of the instrument away from these injection points allow rapid evacuation and ensure flow away from the optical surfaces. The instrument entrance is blocked by the clamshell doors sealing the 1500 Å Al filters and keeping them under vacuum.

The locations of the optical paths and components within the structure are shown in Figure 17. After passing the entrance baffle and filter, radiation is incident on the mirror with the light aperture bulkheads being positioned to absorb most of the stray light. The beam then passes through a slot in the central panel and through the slit/slot mechanism. It is then diffracted by the grating while the zero order is absorbed on the bulkhead ahead of the MHC box (blue in Figure 17). The separately diffracted short- and long-wavelength beams pass through a baffle in the particle radiation shield before hitting the surface of the CCDs. Calibration Light Emitting Diode (LED) sources are mounted inside the particle radiation shield.

The outgassing of the CFRP used for the structure was measured by Outgassing Services International. A test sample was subjected to vacuum bake at 80 °C for 72 h and at 40 °C for 67 h. Using a TQCM, outgassing was measured at collection temperatures of 80 K, 213 K, 253 K and 293 K. The 80 K measurement is the reference for all species condensable down to 80 K. The other measurements show negligible outgassing for species of higher molecular weight (see Table 5).

Data for a temperature of 80 °C are given in the table. Extensive outgassing beyond 72 h was required to reach the desired fluence rate in the flight structure. The main outgassing species was water ($>$99%). Other materials that outgas were used during the manufacture and assembly, *e.g.*, Hysol 9395, Eccobond 285 and Scotch-Weld 1838. Where practical the adhesives were put under vacuum after careful mixing to release all excess gas before application and curing.

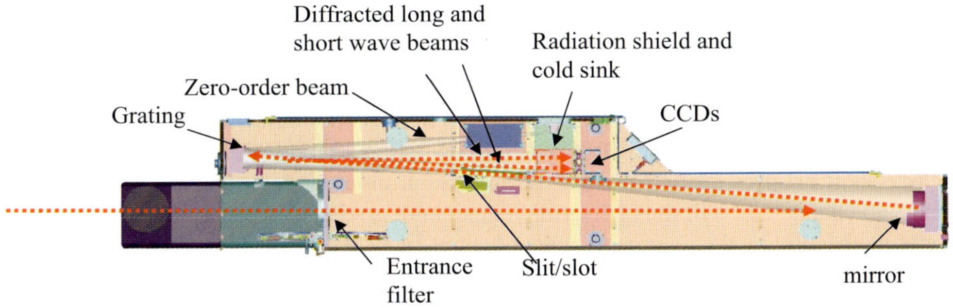

Figure 17 Optical path — view from above the instrument.

Table 5 Accumulated depositions after 72 h with sample at 80 °C.

$T_{collector}$		[µg cm^{-2}]	[%]	% of TML
80 K	TML[a]	288.421	0.04610	100.0
213 K	VCM[b]	2.128	0.00034	0.7
253 K	VCM	0.292	0.00005	0.1
293 K	VCM	0.008	<0.00001	0.0

[a]Total Mass Loss.

[b]Volatile Condensable Materials.

The EIS instrument was thoroughly tested. This was done first at component level and then using a full Mechanical Thermal Model (MTM) of the instrument. The MTM served to verify that the design met (a) the main mechanical requirements of stiffness and strength, (b) the main thermal requirements of thermal balance and stability and (c) stringent cleanliness requirements during manufacture, assembly, test and ability to survive limited periods of exposure to the environment.

The MTM structure was flight-like. The flight electronic and optical subsystems were represented by heaters and dummy masses. To verify the contamination control procedures, the MTM was treated exactly as the FM model for every step of manufacture, assembly and testing. Cleanliness was monitored with QCM sensors and witness mirrors which formed an integral part of the instrument.

The MTM was vibration tested at the UK Aldermaston facility since the mass of the vibration adaptor with the instrument mounted on its semi-kinematic suspension exceeded 500 kg. The test was limited at a frequency of 200 Hz. The higher frequency range was tested separately in an acoustic test. Strength was verified with sine bursts (5 cycles) at 12.5 g in the lateral and 19 g in the longitudinal directions. Half Sine Shock 16.7 $g_n$pk $\times$ 10 mS was performed in the longitudinal direction. All units mounted inside the structure were tested and qualified separately, based on the responses measured during the MTM testing. The first resonance measured was located at 59 Hz. This was only 1 Hz below the requirement and was therefore deemed acceptable.

At the end of the successful MTM test campaign, the cleanliness measurements showed that the rigorous cleanliness regime would result in an overall accumulated contamination for the flight instrument well below the required limits. Except during vacuum testing, the instrument was bagged and purged with dry nitrogen throughout the testing and integration phase. The bagging will be removed at the launch. However, purging with ultra clean nitrogen will continue when the spacecraft is mounted inside the fairing of the launch vehicle up to about two hours before the launch time.

5.2. Thermal Design

The Hinode spacecraft is in a 680 km Sun-synchronous polar orbit with the EIS instrument sun-pointed during normal operations. The instrument temperature requirements are summarized in Table 6. The temperature gradient in the entire base structure that forms the optical bench must be less than 10 °C when in operational mode. The spacecraft orbit will include short eclipses every 8 months, during which time the solar loading on the instrument will change and will thus affect the temperature gradients. In operational mode there is internal heating by the instrument units of about 14 W. This dissipation is almost equally spread between the MHC and the ROE units, but there is also a small dissipation of 0.02 W at the CCDs themselves.

A thermal design using Multi-Layer Insulation (MLI), heaters and radiators has been developed. This design and the related computer models have been validated with data from

Table 6 Summary of temperature requirements (°C).

Item	Survival		Operational	
	Lower	Upper	Lower	Upper
CCD	−100	60	−90	−40
Mirror	−10	40	−10	30
Grating	−10	40	10	30
Filter foil	−200	200	−55	150
MHC unit	−35	65	0	50
ROE unit	−45	65	10	40
Optical bench	−40	40	10	35
Other structure	−90	120	−80	90
HOP[a] actuators	−120	70	−60	70
Slit-slot mechanism	−35	40	10	30

[a]High output paraffin actuator.

thermal testing of the EIS. These models will be used to support EIS on-orbit activities. The instrument is thermally isolated from the spacecraft by low-conductance titanium mounts. Much of the instrument is insulated by MLI blankets having a black Kapton outer layer. The shroud and aperture of the instrument are left uncovered as is a 0.057 m^2 area of the instrument structure adjacent to the MHC unit, which was required to radiate excess heat dissipated by the electronics. Dedicated white-painted radiators are used for cooling the CCDs and the nearby ROE unit. The CCD radiator is connected to a particle shield, which is in turn attached to the CCDs via a strap made of stacked thin copper foils. Thus the particle shield remains colder than the CCDs, protecting them from contamination and radiated heat loads. The mass of the CCD cooling system was kept as high as practical to provide sufficient thermal inertia to minimize temperature variation around the orbit. The CCD radiator is 0.216 m^2 and is fitted to the outward-facing surface of the instrument by eight low-conductance A-frame legs made from Torlon. A shade is placed on the sunward side of the CCD radiator to prevent direct solar heating.

The instrument is fitted with both survival and operational heater circuits. The thermostatically controlled survival heaters are positioned to keep the mirror, grating, and CCDs within survival temperature limits when the instrument is non-operational. There are 12 operational heaters fitted to the optical bench which are used to maintain structure temperatures and gradients during operation. The maximum power budget for these operational heaters is 15 W. In addition there are decontamination or bake-out heaters near the CCDs and make-up heaters near the ROE and MHC units.

Two sets of temperature sensors are fitted to the instrument. The first set comprises 10 sensors and can be monitored by the spacecraft when the instrument is switched off. These show temperature status of critical items and provide the feedback mechanism for control of the survival heaters. The second set comprises about 30 sensors which are monitored by the instrument's MHC unit. A number of these are fitted to the optical bench to ensure a good knowledge of its temperature gradient.

Thermal design cases are derived by taking the extremes of the cold and hot parameters that the instrument may experience during the mission. The cold and hot thermal design cases for operational modes are summarized in Table 7. The hot case assumes the high solar loads experienced at winter solstice together with the degraded end-of-life (EOL) properties.

Table 8 presents a summary of the orbital temperature predictions for the operational cases. In the cold case, the optical bench falls below temperature requirements if heaters are

Table 7 Summary of cold and hot operational thermal analysis cases.

Description	Cold	Hot
Spacecraft interface temperature [°C]	−10	30
Solar flux [$\mathrm{W\,m^{-2}}$]	1290	1421
Earth temperature [K]	248	260
Declination/deg	23.5	−23.5
Earth albedo	0.35	0.25
Beta angle[a]	56.8	71
Radiative properties[b]	BOL	EOL

[a]Angle between sun-vector and satellite orbit plane.

[b]BOL and EOL are beginning-of-life and end-of-life respectively.

Table 8 Temperature predictions for cold and hot operational cases.

Sub-system	Predictions [°C]	
	Cold	Hot
CCD	−58 to −55	−49 to −45
Mirror	19	18
Grating	15	15
Filter foil	6 to 124	134
MHC unit	18	21
ROE unit	29	30
Optical bench	15 to 22	15 to 22
HOP actuators	15 to 19	18 to 33
Slit-slot mechanism	15	16

Note: Range given if T varies by $>2\,°C$ around orbit or across an item.

not used. Therefore the operational heaters are required to warm the optical bench to $\approx 15\,°C$ and to ensure that the gradient requirement is achieved. In the hot case, the optical bench is predicted to be at the lower end of the operational temperature range without heating. However, some heating is required to ensure that the gradient requirement is achieved. Optical bench heater dissipation is predicted to be ≈ 12 W in the cold case and ≈ 4 W in the hot case. Thus there is good margin with respect to the maximum allowable heater dissipation of 15 W.

6. Electronic Design

6.1. Mechanism and Heater Control

The EIS MHC unit consists of the Mechanism Driver Electronics (MDE) and the power subsystem. The Naval Research Laboratory produced the MDE and MSSL produced the MHC back plane, housing and power subsystem. The MHC system provides control of the grating focus mechanism, mirror coarse and fine motion mechanisms, shutter mechanism, slit/slot mechanism, twelve EIS operational heaters, and the two High Output Paraffin (HOP) actuators, each with redundant heater circuits, for opening the Clamshell doors. In addition, the MHC electronics drives two calibration LEDs, two Quartz Crystal Monitors (QCM) for contamination measurement, structure thermistors, an Entran vacuum gauge to check Clamshell pressure, Clamshell door encoders, and other housekeeping sensors within the instrument. The MHC is housed in a single aluminum alloy enclosure (Figure 18).

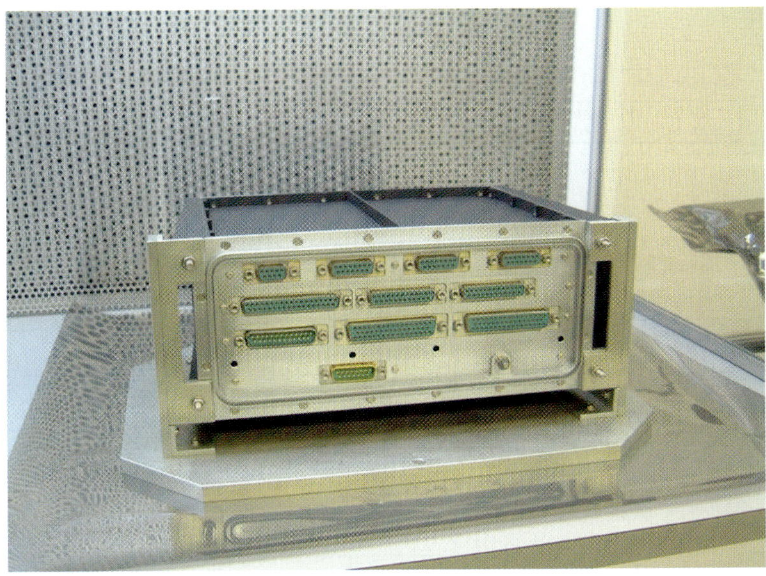

Figure 18 The MHC electronics box is housed in a single enclosure. Boards mounted in the box include the Digital Board (top row of connectors), Analog Board (second row), Auxiliary Board (third row) and the Power Converter Board (bottom connector).

The MDE includes the digital Printed Circuit Board (PCB), the analog PCB and the auxiliary PCB. A block diagram of the MDE is shown in Figure 19.

The digital PCB provides the interface to the ICU via a full duplex, 9600 byte/s, asynchronous RS-422 serial data interface. The control logic is driven by an 8085 radiation hardened microprocessor which also provides the serial input/output functions. The MHC software resides in a 32 k × 8 bit PROM which operates from MHC software uploaded from the ICU to a 64 k × 8 bit RAM device, switchable under software control. The digital PCB provides stepper motor logic for the grating focus mechanism, mirror coarse motion mechanism, and the slit/slot mechanism. It also provides the control logic for the shutter brushless DC motor. Stepper motor voltages are switched and passed to the grating, mirror, and the slit/slot mechanisms.

The analog PCB provides the analog to digital conversion functions, the mirror fine motion control, coarse mirror and slit/slot resolver to digital conversion, voltage and current monitoring, the Entran vacuum gauge operation and structure temperature monitoring. The mirror fine motion control is handled through a PZT drive circuit and a closed-loop strain gauge feedback loop.

The auxiliary PCB provides the operational heater and HOP control, feedback from the optical encoders on the inner and outer clamshell doors, a constant current source for the two calibration LEDs, and switched power to the QCMs and QCM heaters.

The MHC power subsystem provides appropriate electronics, motor, heater and PZT voltages to the MDE. The MHC power interface consists of three +28 V supplies: an electronic supply which provides MHC internal power and shutter power via the MHC power converter, a heater supply which powers structure heaters directly, and a motor supply which powers stepper motors and paraffin actuators. The MHC internal power is generated by two DC/DC converters. Two converters are used to decouple any noise generated on the digital

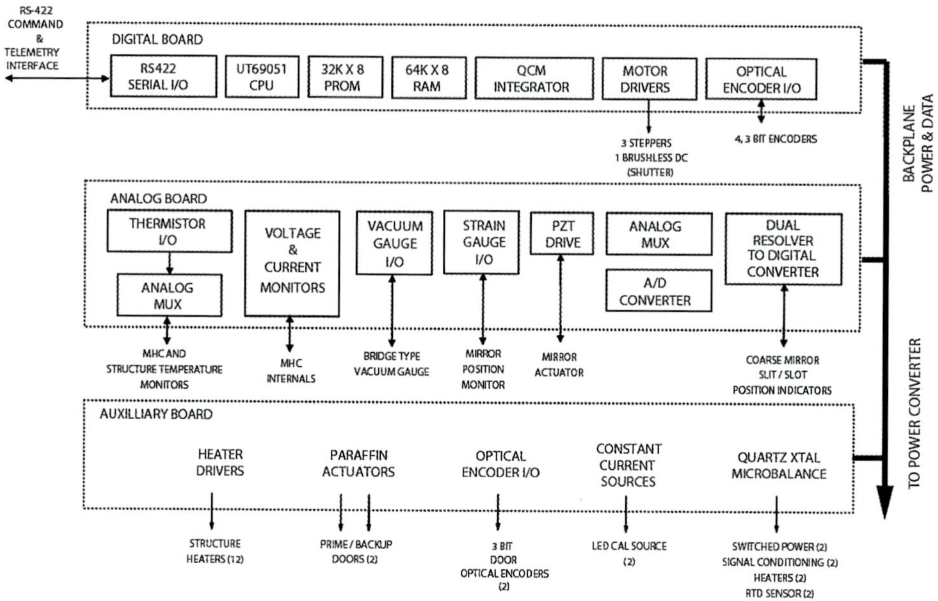

Figure 19 Simplified block diagram of the Mechanism Driver.

and mechanism supplies from the low noise analog supplies, thus enhancing the performance of the analog circuits.

The MHC responds to any of the 35 pre-defined commands issued from the ICU via the RS-422 data link. There are two types of response from the MHC to a command: (a) commands that require a data response (memory dump, housekeeping, *etc.*) will be acknowledged by the return of data if successful, or a Not Acknowledged (NAK) signal in case of failure; (b) commands not requiring a return of data will be recognized by a brief Acknowledged (ACK) signal on successful command completion or a NAK signal for command failure. The MHC does not initiate any data transfer. Commands are not queued by the MHC and a command received while a prior command is pending will generate an error (NAK). The only commands that will be executed while another command is being processed are the ABORT, RESET, and SAFE commands. There are 125 parameters associated with the MHC (heater duty cycles, auto safe parameters, mirror and slit/slot resolver values, *etc.*). Embedded in the flight software is a set of default values assigned to each of these 125 parameters. Using the PARAMETER_SET command, each of the parameters can be changed in-orbit to values within a pre-determined range.

The MHC has three operating modes: Safe, Idle, or Command Active. Following power-up, the MHC is put in the Safe operating mode. In this mode, use of the shutter, coarse mirror, grating, and slit-slot mechanisms are prohibited and any command that attempts to drive these mechanisms will be rejected, *i.e.*, a NAK response will result. All other commands will be accepted and processed normally. The Safe operating mode can also be entered either by command or autonomously. Either a SAFE command or a RESET command will cause the MHC to enter the Safe operating mode. To enter the Idle operating mode, the MOTOR_ENABLE command is issued by the ICU. The Idle operating mode is the MHC state from which all commands are accepted. The Command Active operating mode is entered following the receipt of a valid command from either the Safe or Idle modes. When the MHC is in the Command Active mode, only the ABORT, RESET, or SAFE commands are accepted, all others are rejected. The MHC will transition from the Command Active mode to the Idle or Safe mode following completion of the current command.

The MHC supports two memory modes; Programmable Read Only Memory (PROM) mode and Random Access Memory (RAM) mode. In PROM mode, the baseline code residing in a PROM is executed. In RAM mode, the program currently loaded in a RAM is executed. The baseline code is copied from PROM to RAM at power-up. On power-up, the default is PROM mode. Switching between modes is accomplished via the MEMORY_MODE command and always results in a return to the SAFE operating mode. Updated MHC operating code can be uplinked from the ICU to the MHC RAM using the MEMORY_UPLOAD command.

6.2. On-Board Data Processing Unit

The ICU is an on-board processor that controls the entire EIS instrument. Located in the spacecraft bus section, it also handles interfacing between the EIS spectrometer and the spacecraft. Figure 20 gives an overview of the ICU electronics. The circuitry is located on five printed circuit boards: spacecraft interface and processor board, camera and mechanism controller board, analog monitor board, power supply unit and a backplane. The spacecraft interface and processor board is based around a TEMIC 21020 Digital Signal Processor (DSP) running at a clock speed of 20 MHz. There are two FPGAs, a Static RAM (SRAM; 1 Mb) and a boot PROM of 8 kb to support the DSP. One FPGA decodes the address space for the memory and I/O. It also implements the Watch Dog and Spacecraft timer functions and a "boot-strap" code power-up function. The second FPGA deals with the spacecraft digital interfaces.

The spacecraft interface is based on three links, command and housekeeping or status, both 62.5 kbps and science mission data transfer at 2 Mbps. Each of the interfaces incorporates First In First Out (FIFO) buffers. The camera and mechanism controller board contains onboard application code storage in 1 Mb of Electrically Eraseable PROM (EEPROM) which can hold two versions of the code, CCD image buffers in 4 Mb of SRAM for the raw science data and two further FPGAs. One FPGA handles the 32 Mbps high speed link between the ROE and the ICU while the second controls the RS422 9.6 kbps interfaces between the ICU and the ROE and MHC. The analog monitor contains the primary power interface and current limiter for the instrument. It is responsible for temperature, current and voltage monitoring, primary and secondary line switching and heater switching. FET switches are used to switch primary power to the MHC, bake-out and substitution heaters and secondary unregulated power to the ROE. An FPGA is used for control of the monitoring function and bus interface along with an ADC, analog multiplexers and operational amplifiers. Secondary power line conditioning is provided by the power supply unit board. There are regulated power lines for the ICU and unregulated ones for the ROE. The backplane provides the interface connections between the above boards.

Buffer logic on the daughter boards for the backplane is included in the FPGAs but is not shown in the diagram. The bootstrap code in PROM is written in assembly language and supports the loading and dumping of the operational code from either bank of the EEPROM or spacecraft to RAM, as commanded. In order to facilitate the software development the operational code is designed to be modular. This code is written in C operating under the

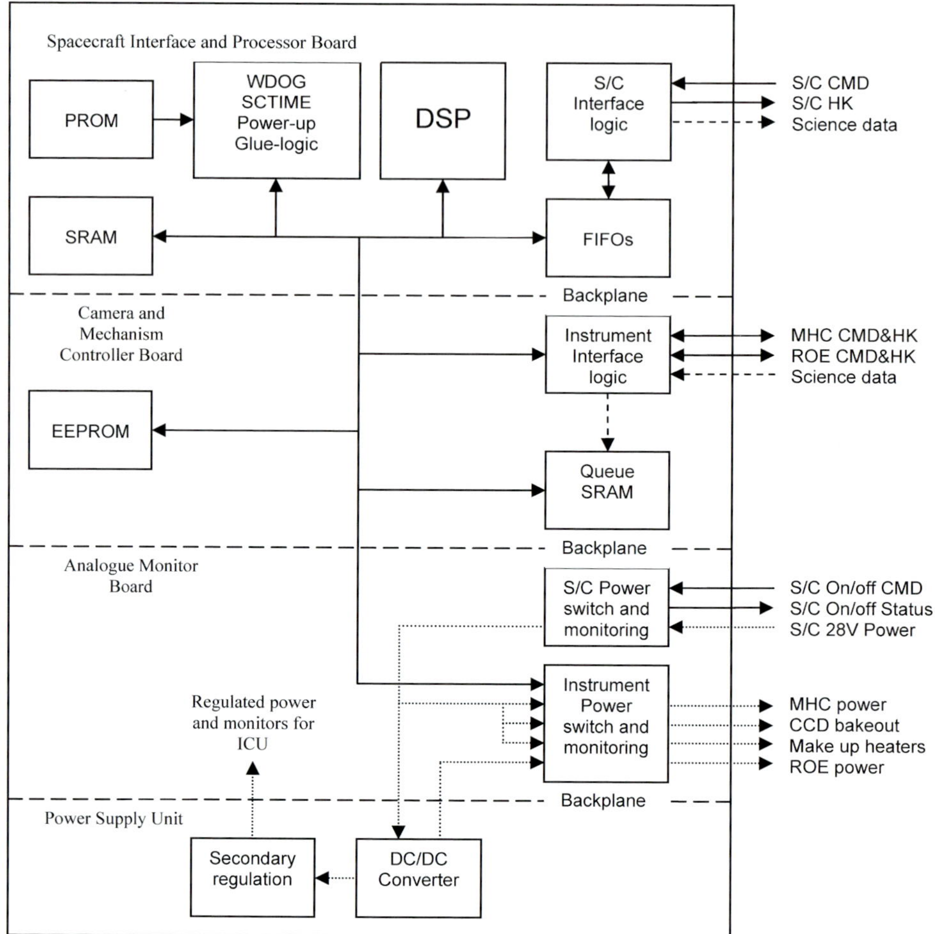

Figure 20   Block diagram for the EIS Instrument Control Unit.

Virtuoso real-time operating system (Wind River Inc.). This is necessary to support the spacecraft and instrument requirements for task scheduling and inter-task communication.

An overview of the EIS software modes is given in Figure 21. Boot mode is used to load, dump and run the main application programme. The boot code also supports housekeeping requests. Standby mode is the first mode entered after invoking the application code. In this mode only the ICU and "make up" heaters in the ROE and MHC are powered. In manual mode the ROE and MHC are powered and the instrument can be fully configured ready for science operations. Science sequences are loaded into memory from the ground and the next sequence to be run is selected. In auto mode, sequences are run together to form a complete science observation or study. When a study is complete, the instrument can be commanded back to manual mode and the first sequence of the next study selected.

Bake-out mode is used to decontaminate the CCDs. A closed loop heater controller warms the CCDs in a controlled fashion to the bake-out temperature where the selected value is held. Likewise the CCDs are allowed to cool at a controlled rate after the bake-out is complete. From any of the above application code modes, emergency mode can be en-

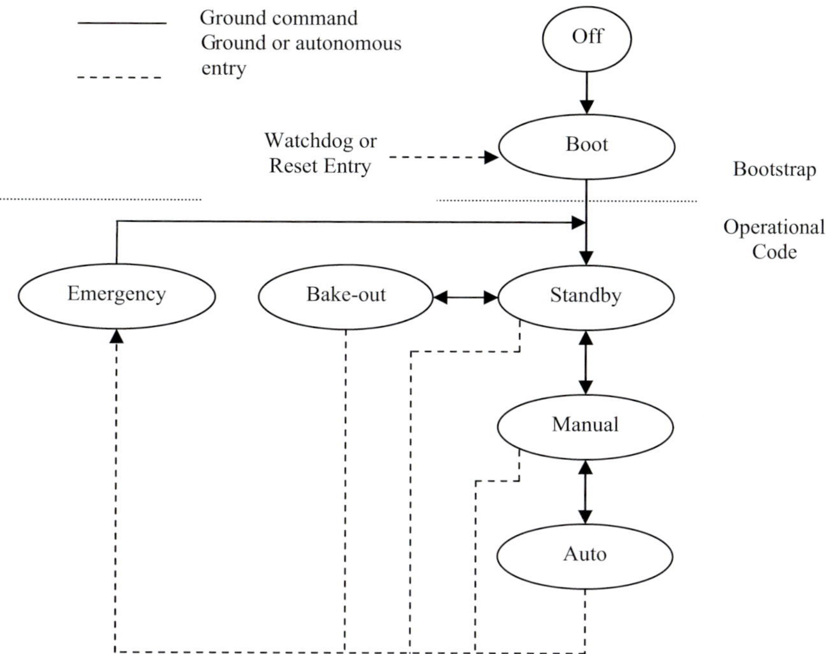

Figure 21 Diagram showing the operation of the EIS on-board software.

tered either by command or by the instrument health monitor software module. The health monitor checks for two consecutive values of a parameter exceeding a defined limit. Parameter limits are loaded from the ground, except for default values for the maximum CCD temperatures. The parameters checked are currents, voltages and temperatures. Emergency mode can also be entered if communication with the MDP or MHC is lost for any reason.

7. Instrument Calibration and Performance

The laboratory calibration of EIS follows the successful example of CDS (Lang *et al.*, 2000). The instrument was calibrated just prior to final hardware delivery. The EIS entrance aperture was illuminated with EUV radiation and a series of calibration images obtained. The calibration comprised an end-to-end test of the entire EIS instrument.

Two EUV light sources were used. One was a Penning discharge lamp (Finley *et al.*, 1979, Berkeley Photonics Inc.) which, at the focus of a custom-built collimator illuminated fully the EIS aperture. This laboratory light source was used to focus the spectrometer and to obtain the wavelength calibration. The other source was a secondary radiometric standard (Hollandt *et al.*, 2002), a hollow-cathode lamp combined with a collimating telescope, previously calibrated against the electron storage ring BESSY I as a primary standard of calculable synchrotron radiation (Ulm and Wende, 1997). This source beam was 5 mm in diameter and did not fill the EIS aperture. The instrument aperture response was computed by obtaining successive exposures over the aperture. The overall effective area of the EIS instrument was a weighted average of the efficiency. The laboratory calibration of EIS is described in detail elsewhere (Lang *et al.*, 2006) so only a brief account is given here. The calculated instrument response to three typical solar coronal spectra is also presented.

7.1. Wavelength Calibration

The instrument was fully assembled at the Rutherford Appleton Laboratory in the UK and aligned in its flight configuration. Tests with the Penning EUV discharge lamp were performed with the instrument in a high vacuum facility. A 150 mm diameter multilayer coated spherical mirror was used to collimate the radiation from the Penning lamp and fill the EIS telescope aperture to permit optimal focusing. The lamp was operated with He or Ne gas and Mg electrodes. Representative spectra recorded with a 10 μm slit are shown in Figure 22 for the long wavelength band using neon. Figure 23 shows representative spectra from the short wavelength band from ionized magnesium and neon. Figure 24 shows a He II 256 Å line profile with a Gaussian fit also indicated for both linear and log scales. The FWHM He line width is 0.056 Å or 2.5 pixels for a measured spectrometer resolving power of $\lambda/\Delta\lambda = 4570$ at this wavelength. Many of the spectral lines observed are found in the NIST database (2005) and in Kelly's compilation (Kelly, 1987), but a number of new lines of Ne II, III, and IV were identified in the course of this work (Kramida *et al.*, 2006; A.E. Kramida, 2006, private communication; C.M. Brown, 2006, private communication), a tribute to the high resolution and sensitivity of EIS. One minute (long wavelength band) and 5 minute (short wavelength band) exposures were recorded using the 1 arc sec slit.

Spectral images were formed in the upper half of each CCD, just above the midline. For these images 50 rows were averaged to produce a spectral intensity curve. These intensity

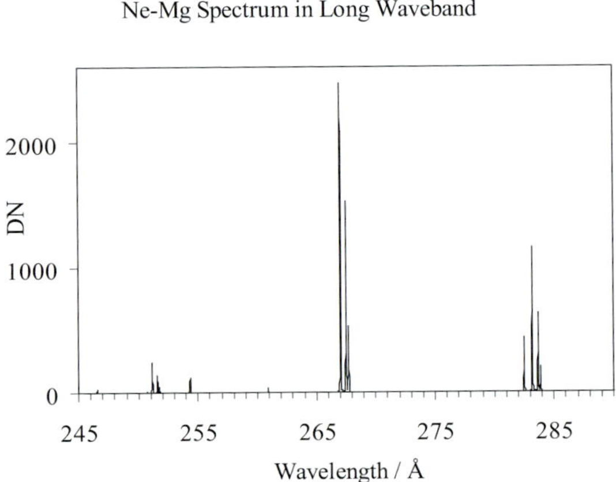

Figure 22 Neon and Magnesium long λ Penning discharge spectrum recorded with the EIS spectrometer.

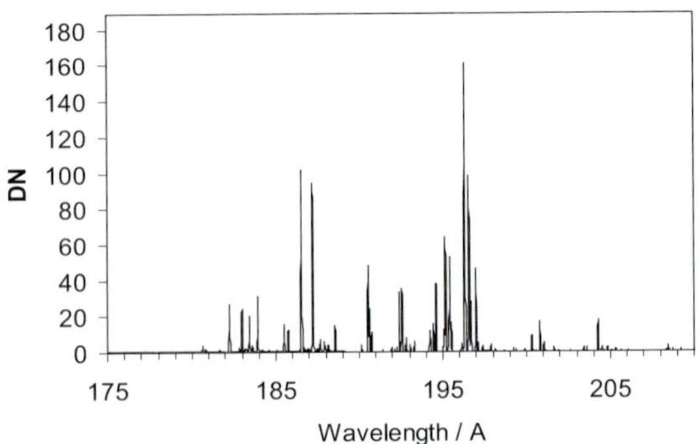

Figure 23 Representative short λ Ne-Mg spectra recorded with the EIS spectrometer.

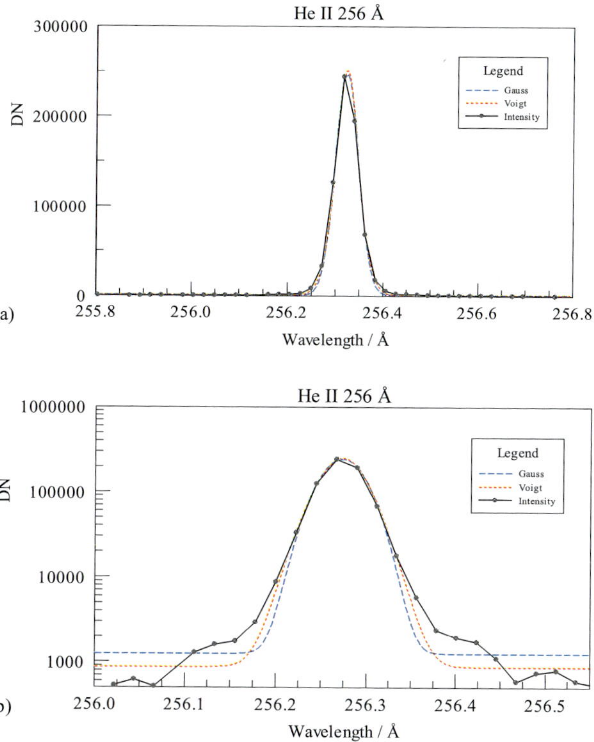

Figure 24 Profile of the He II 256.32 Å spectral line from a Penning discharge lamp, as recorded by EIS on (a) linear and (b) semi-log scales. Also shown are fitted Gaussian (blue) and Voigt (red) line profiles. The fitted FWHM for the Gaussian profile is 0.056 Å.

curves were measured using the Gfit program (Engström, 1998) to fit Gaussian shapes to the line profiles. Line positions were determined to ≈ 0.1 column accuracy for good lines, and the FWHM of unblended lines was typically 2.5 pixels. The area under the Gaussian curve was also computed by the program.

This list of measured lines was then compared with a list of standard lines and fitted with the lowest order polynomial practical. In general, no improvement in fit was found for orders higher than second. The dispersion function for the long wavelength band results from a second order polynomial fit to the data from 32 standard lines and is given by

$$\lambda(p) = \lambda_0 + Ap + Bp^2 \tag{1}$$

where p is the pixel (or column) number and λ_0, A, and B are the polynomial coefficients. Here, $\lambda_0 = 199.9389$ Å, with $A = 0.022332$ and $B = -1.329 \times 10^{-8}$ while the standard deviation was 0.00415 Å. λ_0 can be interpreted as the wavelength of the edge of column 0 for each detector. A is the linear term in Å/pixel. Figure 25 shows a plot of the reference wavelength minus the fitted wavelength for this detector. Likewise, for the short wavelength band using the format of Equation (1), the polynomial coefficients are: $\lambda_0 = 166.131$ Å, $A = 0.022317$ and $B = -1.268 \times 10^{-8}$. The standard deviation of the fit was 0.00386 Å. Figure 26 shows a plot of the deviations of the fitted wavelengths from the standard wavelengths of 65 lines.

Figure 25 Wavelength
deviations for standard lines; EIS
long wave band.

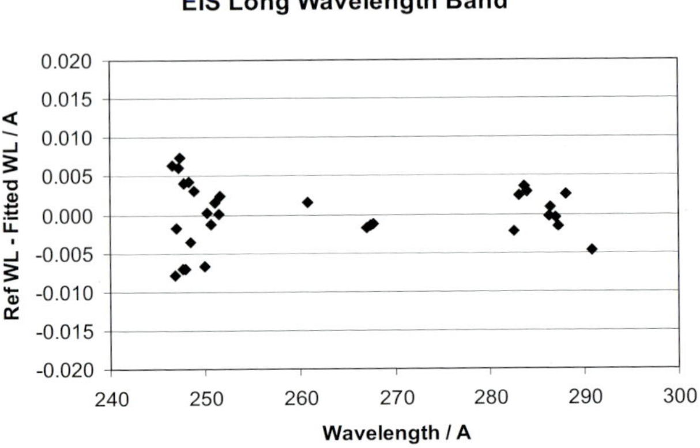

Figure 26 Wavelength
deviations for standard lines; EIS
short wave band.

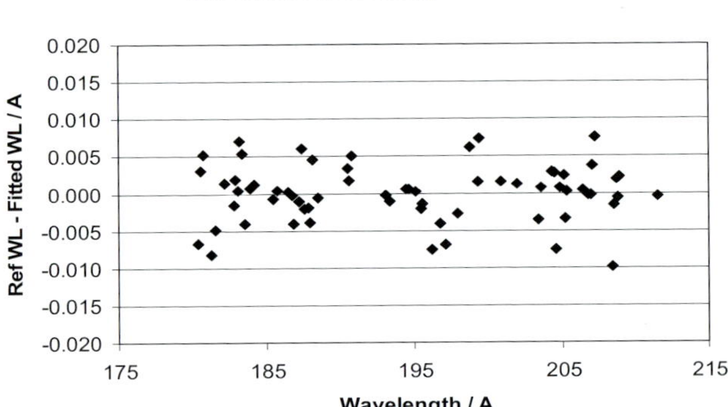

7.2. Effective Aperture Determination

The geometric aperture of the mirror and grating combination may be calculated from accurate mechanical drawings. In addition, allowance must be made for the obstructions associated with the entrance filter structure and filter support frame. The vignetting of the long wavelength (LW) CCD by the slit/shutter housing must also be calculated. This housing causes wavelength dependent vignetting of the radiation from grating to the detector for wavelengths greater than 272 Å and was measured during the radiometric calibration. The housing also vignettes the beam incident on the grating for some mirror positions. The mirror aperture is shown as a function of coarse mirror position in Figure 27. Details are given by Lang *et al.* (2006).

For measurements in the vacuum chamber, the calibrated high current hollow cathode lamp (Danzmann *et al.*, 1988) emits unpolarized line radiation from the carrier gas (Ne or He) and from sputtered cathode material (99.5% Al). The lamp illuminated a pinhole at the focus of a Wolter type II telescope while a 5 mm aperture stop placed just after the telescope mirror defined the collimated output beam. This telescope polarizes the output beam. The output of the source is the measured sum of the different polarization outputs (Hollandt *et al.*, 2002; Hollandt, 1994). However, it is known that polarization effects are not appreciable at normal incidence, *e.g.*, Samson (1967), and as EIS is a normal incidence instrument, polarization effects can be ignored. The aperture was scanned from one edge of

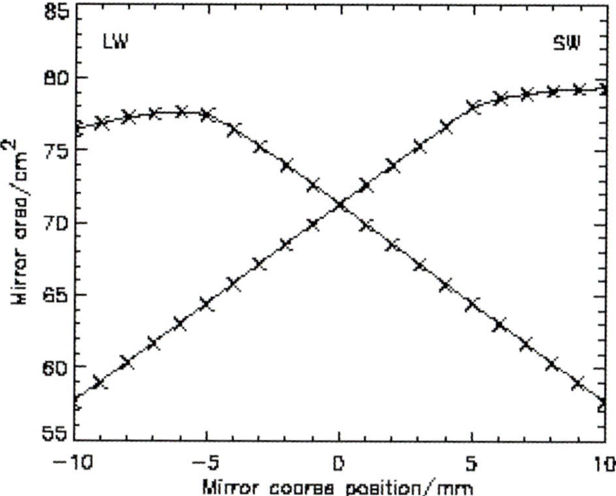

Figure 27 Mirror aperture as a function of coarse mirror position. The small fall-off in area for the LW aperture from 6 mm to 10 mm results from vignetting by the slit/shutter housing.

Table 9 Source calibration line list.

Wavelength [Å]	Spectrum	Photon flux [photons s^{-1}]	Last calibration uncertainty (1σ)	Aging uncertainty (1σ)
204.3 – 208.9	Ne III	3.43×10^5	8%	10%
251.1 – 251.7	Ne III	4.39×10^5	"	"
256.3	He II	8.53×10^6	"	"
276.1 – 267.7	Ne III	1.01×10^6	"	"
282.5 – 283.9	Ne III	2.61×10^6	"	"

the mirror to the other by moving EIS horizontally in 10 mm steps on its translation table between exposures.

After each horizontal scan, the calibrated source was moved up or down nominally by 10 mm and the horizontal scan repeated to map the whole aperture. With helium as the hollow cathode base gas, only the He II 256.3 Å line could be used to illuminate the long wavelength part of the aperture. With neon as the base gas, numerous lines were observed in both the EIS bands and the scan was over the full aperture. The radiometrically calibrated ranges together with the output fluxes and uncertainties are given in Table 9.

Prior to final instrument assembly, mirror and grating reflectivities and entrance and slit filter transmissions were measured at the Brookhaven Synchrotron Light Source (Seely *et al.*, 2004). Measurements for these subsystems are shown in Figures 28 and 29 along with previously estimated values. The flight grating groove depth, the most demanding of the grating specifications to achieve, has a measured value of 67 Å compared with the specified value of 60 ± 4 Å. This small added depth moves the peak groove efficiency to longer wavelength by $\approx$20 Å and slightly affects the resulting response curves. Another factor in the comparison is that a fixed groove efficiency of 35% was used whereas the efficiencies estimated using the achieved groove depth vary between 30% and 40% with wavelength for the short wavelength band.

The measured and estimated transmission of the slit filter, including its supporting mesh, is plotted as a function of wavelength in Figure 29. The previously estimated filter transmis-

Figure 28 Measured and predicted values for the product of flight mirror and grating reflection efficiencies.

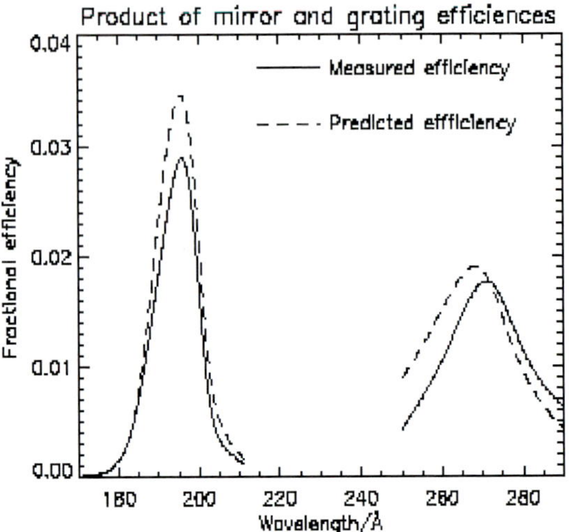

Figure 29 Measured and predicted values for the slit filter transmission and for the product of entrance and slit filter transmissions.

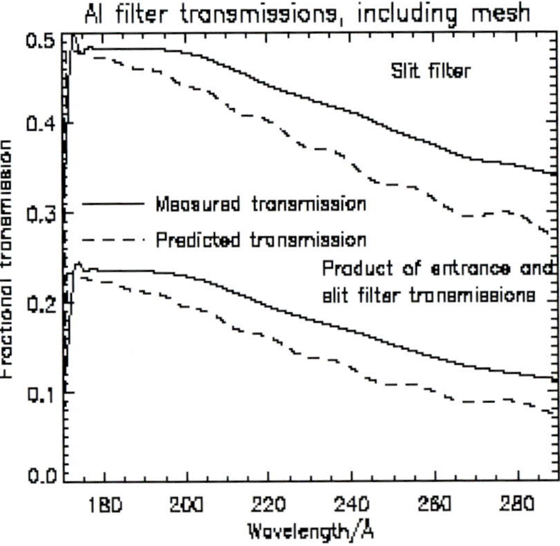

sion was based on a thickness of 1500 Å aluminum with a 75 Å Al_2O_3 layer on each side. The measured filter transmission is best fitted by 1500 Å aluminum with 80 Å Al_2O_3 layers on each side. However, more recently available values of Al and Al_2O_3 optical constants were used for fitting the measured transmissions. The products of the transmissions for entrance and slit filters are also shown in Figure 29. The entrance filter was not included in the end-to-end calibration measurement of the instrument but was fitted to the EIS instrument shortly before the launch of Hinode.

The detector quantum efficiency (QE) used in the original instrument sensitivity estimate presented by Lang, Kent, and Seely (2002) was 0.8 based on measurements on CCDs with "UV enhanced" backside treatments by the CCD manufacturer (Stern *et al.*, 2004). However, the CCDs chosen for EIS did not have the enhanced treatment and are expected to have a 25 to 40% lower QE. The QE values of two engineering quality CCDs of the EIS type were measured using synchrotron radiation and the same apparatus as used to mea-

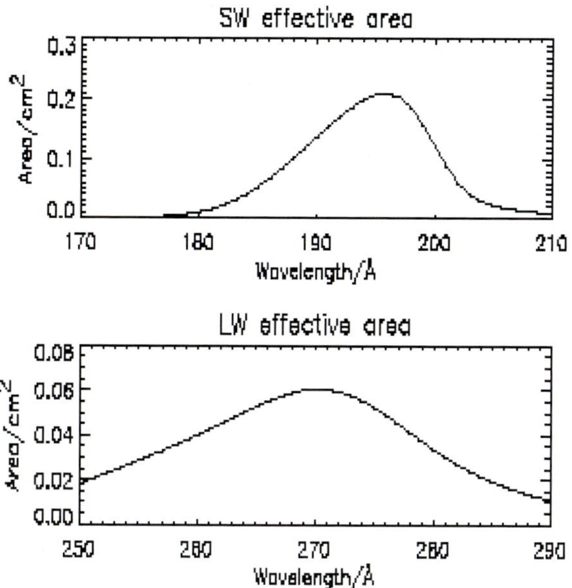

Figure 30 The SW and LW effective areas based on measurements of the efficiencies of the individual optical elements (filters, mirror, grating, CCDs) of EIS. Vignetting in the LW band by the slit and shutter housing is also included.

sure the flight mirror and grating reflectivities. For the short-wavelength band the measured efficiencies averaged 48% and 40% and for the long-wavelength band were 42% and 35% respectively. These CCDs were from the same batch but a different wafer from those chosen for flight. Noting the variation in the measured results, it was decided to use the average of the measurements for the EIS calibration model. The CCD quantum efficiencies adopted were $44 \pm 4\%$ for the short wavelength band and $40 \pm 4\%$, $39 \pm 4\%$ and $37 \pm 4\%$ at the longer wavelengths of 250 Å, 270 Å, and 290 Å, respectively. These efficiencies can be combined to give the predicted effective areas for the two EIS wavelength bands, as shown in Figure 30.

In applying the laboratory calibration to solar observations, the change of illumination from a small source filling a small part of the aperture to full solar illumination must be allowed for as well as the effect of making a coarse movement of the mirror. The conversion from data numbers/sec to radiance (I_λ) can be made using Equation (2)

$$I_\lambda = (D_{\mathrm{L}}/D_{\mathrm{p}})(1/A)(1/A_{\mathrm{s}})\left((180.0 \times 60.0^2)/\pi\right)^2 \text{ photons cm}^{-2}\,\text{s}^{-1}\,\text{sr}^{-1} \qquad (2)$$

where A is the aperture area in cm^2 and A_{s} is the area of the spectrometer slit or slot illuminated by the source in arc sec^2. D_{L} is the digital signal in a spectrum line measured in data numbers at the output of the camera ADC. To relate the data numbers to the number of detected photons, the number of electrons was calculated using the electron-hole pair creation energy of silicon (($3.66 \pm 0{:}03$) eV at room temperature (Scholze *et al.*, 2000) corresponding to ($3.68 \pm 0{:}04$) eV at 215 K (Canali *et al.*, 1972; Sze, 1981), the operating temperature of the CCDs) and the incident photon energy and then converted to data numbers using the gain of the camera (6.60 ± 0.03 electrons per data number).

The factor D_{p}, the responsivity measured during the end-to-end calibration of EIS, is obtained by dividing the measured data number values corresponding to the calibration wavelength ranges by the appropriate calibrated source output as given in Table 9. Predicted responsivities are derived from the effective areas based on the measurements for the individual elements as shown in Figure 30. As the predicted responsivity is based on flight unit results apart from the CCDs where engineering model devices were used for the Brookhaven SLS measurements, the final responsivity for EIS must reconcile the end-to-end with the predicted measurements. The quantities involved for the standard source wavelength intervals are given in Table 10.

The normalizing factor (Norm = 1.60) was chosen to obtain agreement between the measured and predicted responsivity values for $\lambda = 267.25$ Å. The adopted long-wavelength band responsivity is the laboratory result with the normalized predicted responsivity scaled to give the measured responsivity at 251.3 Å and 283.4 Å. For the long-wavelength band, Figure 31 shows the responsivities as measured using the calibrated source (crosses with error bars), the predicted responsivity (dash-dot line), the normalized responsivity (predicted responsivity times 1.60; dotted line) and the adopted laboratory responsivity (solid line). The uncertainty in the laboratory responsivity is taken as the sum of the average uncertainty of the measured responsivities (15.6%) and the uncertainty of the predicted responsivity (13.3%), namely 20.6%, as indicated by the dashed lines in Figure 31.

For the short-wavelength band (see Table 10), the normalized predicted sensitivity is lower than the measured responsivity. The laboratory responsivity is taken as the measured result, extended to other wavelengths by matching the normalized predicted data to the mea-

Table 10 Line data numbers, responsivities and comparison with predicted values.

Band	λ [Å]	Data numbers [DN sec^{-1}]	Responsivity [DN/photon]		Meas./ pred.	Meas./ norm.
			Measured	Predicted		
LW	251.3	$8.54 \pm 1.00 \times 10^2$	$1.95 \pm 0.34 \times 10^{-3}$	$1.47 \pm 0.20 \times 10^{-3}$	1.32	0.83
LW	256.3	$2.77 \pm 0.28 \times 10^4$	$3.25 \pm 0.53 \times 10^{-3}$	$2.24 \pm 0.30 \times 10^{-3}$	1.45	0.91
LW	267.25	$6.70 \pm 0.50 \times 10^3$	$6.63 \pm 0.98 \times 10^{-3}$	$4.15 \pm 0.55 \times 10^{-3}$	1.60	1.00
LW	283.4	$8.30 \pm 0.51 \times 10^3$	$3.18 \pm 0.45 \times 10^{-3}$	$1.66 \pm 0.22 \times 10^{-3}$	1.92	1.20
SW	205.9	$8.95 \pm 0.80 \times 10^2$	$2.61 \pm 0.41 \times 10^{-3}$	$1.31 \pm 0.18 \times 10^{-3}$	1.98	1.24

Figure 31 The measured long wavelength band responsivity compared with predicted values. The points and curves are as described in the text.

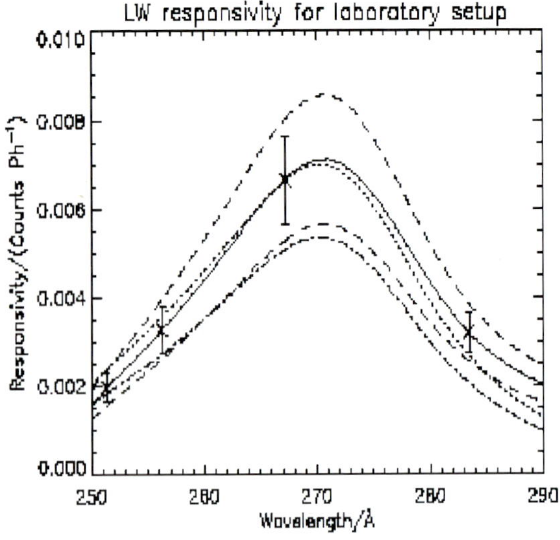

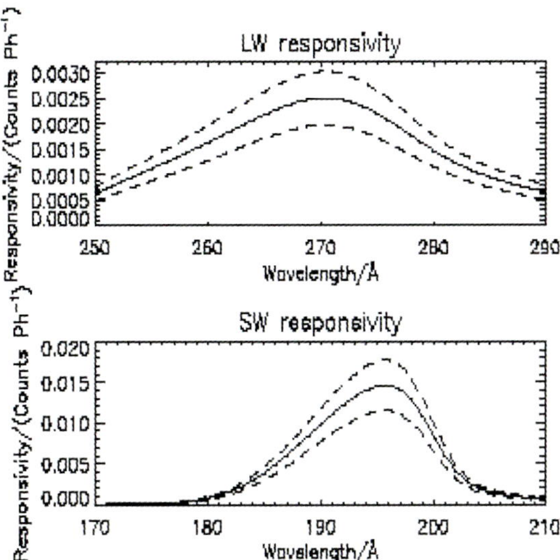

Figure 32 The long and short wavelength responsivities, D_p, as deduced from the calibration measurements. Dashed lines indicate the range of uncertainty.

sured result. Again the uncertainty in the result is taken as the sum of the uncertainties in the measured and predicted responsivities, namely 20.5%. Final measured responsivities are plotted against wavelength in Figure 32 for both the EIS bands and include the front filter transmission which has uncertainty 5%. If the normalizing factor of 1.6 at 267 Å (see Table 10) were ascribed entirely to CCD quantum efficiency, the figure of 39% measured in the Brookhaven CCD sub-system test should be increased to 59%. This is very similar to the QE value of 60% that was measured for the XRT CCDs at this wavelength (Hara, 2003; Sakao *et al.*, 2004). The XRT CCDs have an identical specification to the EIS flight devices.

When additional uncertainties of 2%, 4%, and 4% are added to allow for the uncertainties in the slit area, the mirror area, and for variations due to the spread of measurements of the QE of the engineering model CCDs, respectively, the overall relative standard uncertainty in the absolute responsivity calibration is 22%.

7.3. Instrument Performance

For an ideal optical system with no losses, the number of photons of wavelength λ entering the slit per second in a height interval corresponding to one pixel in the spatial dimension on each CCD, $N_\mathrm{slit}(\lambda)$ is given by

$$N_\mathrm{slit}(\lambda) = \phi_\lambda A\left(a/f^2\right) \tag{3}$$

where ϕ (photons cm^{-2} s^{-1} sr^{-1}) is the intensity of the solar radiation, A (cm^2) is the mirror area, a (cm^2) is the area of the slit corresponding to one spatial CCD pixel in height times the width of slit, and f is the focal length of the mirror. These photons are then imaged onto the CCD in a manner determined by the magnification in the spectrometer section of the instrument. Thus, the number of photons per second registered on a pixel of each CCD is given by

$$N_\lambda = \phi_\lambda A\left(a/f^2\right)(1/n_\mathrm{pix}) \tag{4}$$

where the division by n_pix provides the desired solid angle per detector pixel. The design goal was to have the narrow slit match the pixel size. However, the actual slit widths result

Table 11 Quiet sun count rates (s^{-1} pixel^{-1}).

Ion	λ [Å]	Log T [K]	Incident	Detected	DN
Short wavelength					
Fe X	184.54	6.00	13.09	0.86	2.28
Fe VIII	185.21	5.60	8.87	0.69	1.84
Fe XII	186.88	6.10	13.97	1.60	4.21
Fe XI	188.23	6.10	17.87	2.65	6.91
Fe XII	193.52	6.10	29.06	8.13	20.60
Fe XII	195.12	6.10	48.15	14.58	36.63
Fe XIII	196.54	6.20	6.44	1.95	4.85
Fe XII	196.65	6.10	6.43	1.93	4.82
Fe XIII	202.04	6.20	18.02	1.49	3.60
Long wavelength					
He II[a]	256.32	4.90	18.50	1.04	1.98
Si X	258.37	6.10	13.54	0.88	1.66
Fe XIV	264.79	6.30	17.41	1.65	3.05
Si VII	275.35	5.80	9.35	0.89	1.59
Si VIII	277.06	5.90	8.36	0.71	1.25
Fe XV	284.16	6.30	68.25	2.77	4.78

[a]Intensity is a factor two higher than the CHIANTI value; see text.

in a need for small corrections. For the narrow slit (nominally 1 arc sec wide), the correction factor, n_{pix} is 1.067 and 1.087 for detectors A and B, respectively. The corresponding values for the 2 arc sec slit are 2.080 and 2.119. Note that the combination of factors results in the same solid angle per detector pixel, ω_{d}.

In practice, the actual number of registered photons will be reduced by additional factors due to the transmission of the two aluminum filters, the reflectivity of the mirror, the efficiency of the grating, and the quantum efficiency of the detectors (see Figure 1). Thus, the basic expression for the number of photons registered in each detector pixel per second is

$$N_\lambda = \phi_\lambda A \omega_{\mathrm{d}} T_{\mathrm{ff}}(\lambda) T_{\mathrm{spider}} R_{\mathrm{m}}(\lambda) E_{\mathrm{g}}(\lambda) V_{\mathrm{d}}(\lambda) T_{\mathrm{rf}}(\lambda) E_{\mathrm{det}}(\lambda) \tag{5}$$

where $T_{\mathrm{ff}}(\lambda)$ and $T_{\mathrm{rf}}(\lambda)$ are the transmissions of the aluminum front filter assembly and spectrometer entrance filter respectively; T_{spider} is the fractional area of the front filter assembly that is blocked by supporting structural members; $R_{\mathrm{m}}(\lambda)$ is the reflectivity of the mirror coatings; $E_{\mathrm{g}}(\lambda)$ is the grating efficiency which includes both the groove efficiency and the reflectivity of the multi-layer coatings, $V_{\mathrm{d}}(\lambda)$ is a vignetting factor to account for the vignetting at the long-wavelength end of the long-wavelength detector and $E_{\mathrm{det}}(\lambda)$ is the detector quantum efficiency.

The response of the EIS instrument to a range of solar conditions has been calculated using synthetic spectra from the CHIANTI data base (Version 4: Dere *et al.*, 1997; Young *et al.*, 2003). For line emission from an optically thin plasma, the intensity is

$$I_\lambda = \int G(T)\mathrm{DEM}(T)\,\mathrm{d}T \text{ photons cm}^{-2}\,\mathrm{s}^{-1}\,\mathrm{sr}^{-1} \tag{6}$$

Table 12 Active region count rates (s^{-1} pixel^{-1}).

Ion	λ [Å]	Log T [K]	Incident	Detected	DN
Short wavelength					
Fe X	184.54	6.00	147.56	9.68	25.71
Fe XII	186.85	6.10	186.61	21.29	55.86
Fe XII	186.88	6.10	253.87	29.16	76.49
Fe XI	188.23	6.10	261.35	38.80	101.04
Fe XI	188.30	6.10	99.28	14.91	38.82
Fe XII	192.39	6.10	206.05	52.70	134.26
Fe XI	192.83	6.10	55.93	14.85	37.75
Fe XII	193.52	6.10	548.26	153.43	388.63
Ca XIV	193.87	6.50	24.62	7.05	17.82
Fe XII	195.12	6.10	907.96	274.93	690.66
Fe XII	195.13	6.10	89.64	27.15	68.20
Fe XIII	196.54	6.20	147.64	44.60	111.24
Fe XII	196.65	6.10	116.40	35.03	87.31
Fe XIII	197.43	6.20	48.21	13.82	34.30
Fe XIII	200.02	6.20	180.32	31.27	76.62
Fe XIII	201.13	6.20	203.14	24.05	58.60
Fe XIII	202.04	6.20	439.19	36.18	87.78
Fe XIII	203.83	6.20	749.17	32.72	78.68
Long wavelength					
He II[a]	256.32	4.90	58.38	3.26	6.24
S XIII	256.68	6.40	143.76	8.26	15.78
Fe XIV	257.39	6.30	158.04	9.56	18.20
Si X	258.37	6.10	249.17	16.11	30.56
Fe XVI	262.98	6.40	154.60	13.29	24.78
Fe XIV	264.79	6.30	543.86	51.43	95.20
Fe XIV	270.52	6.30	250.65	27.82	50.41
Fe XIV	274.20	6.30	396.32	40.28	72.01
Fe XV	284.16	6.30	3229.59	131.12	226.17

[a]Intensity is a factor two higher than the CHIANTI value; see text.

where $G(T)$ is the emissivity function but includes the temperature-independent parameters and DEM (T) is the differential emission measure function. Using CHIANTI V4 data and assuming a constant pressure of 10^{16} cm^{-3} K, $G(T)$ was calculated for all emission lines in the EIS wavelength bands. The $G(T)$ data and the CHIANTI DEM curves for quiet Sun, active region and flare cases were then used to compute spectra.

Using the synthetic spectra and the effective areas, we can calculate the number of photons registered in each detector pixel per second as

$$N_\lambda = I_\lambda A_{\text{eff}}(\lambda)\omega_{\text{d}} \tag{7}$$

where $A_{\text{eff}}(\lambda)$ is the effective area which includes all of the factors from Equation (5) except for ω_{d} which is the solid angle per pixel. Use of the appropriate values of D_{p}, then allows the

Table 13 Flare count rates ($\mathrm{s}^{-1}\,\mathrm{pixel}^{-1}$).

Ion	λ [Å]	Log T [K]	Incident	Detected	DN
Short wavelength					
Fe XXI	187.92	7.00	34298.14	4822.71	12579.11
Fe XI	188.23	6.10	9368.79	1391.00	3622.25
Fe XXIV	192.03	7.20	189315.64	46799.19	119458.12
Fe XII	192.39	6.10	7626.77	1950.48	4969.30
Ca XVII	192.82	6.70	219452.22	58220.96	148003.11
Fe XII	193.52	6.10	20345.04	5693.67	14421.38
Ca XIV	193.87	6.50	8714.48	2494.83	6307.85
Fe XII	195.12	6.10	33682.56	10198.99	25621.28
Fe XIII	196.54	6.20	6142.33	1855.63	4627.89
Fe XII	196.65	6.10	4241.22	1276.29	3181.27
Fe XIII	200.02	6.20	7737.11	1341.60	3287.67
Ca XV	200.97	6.60	14209.02	1783.61	4350.19
Fe XX	201.05	7.00	12026.51	1468.70	3580.81
Fe XIII	202.04	6.20	19404.28	1598.67	3878.44
Fe XIII	203.83	6.20	32127.88	1403.17	3374.33
Fe XVII	204.65	6.70	61496.85	2164.85	5185.02
Long wavelength					
Fe XVI	251.06	6.40	66188.77	2345.63	4579.51
Fe XXII	253.17	7.10	32180.93	1399.04	2708.70
Fe XVII	254.87	6.70	66730.63	3344.69	6432.54
Fe XXIV	255.11	7.20	101716.30	5198.91	9989.00
He II	256.32	4.90	108063.31	6049.24	11568.22
S XIII	256.68	6.40	52808.81	3035.89	5797.33
Fe XVI	262.98	6.40	116130.12	9985.35	18611.88
Fe XXIII	263.77	7.10	91040.15	8175.54	15192.91
Fe XIV	264.79	6.30	31941.00	3020.41	5591.22
Fe XVII	269.41	6.70	14806.13	1631.17	2967.73
Fe XIV	270.52	6.30	15148.67	1681.29	3046.37
Fe XXI	270.57	7.00	29343.42	3256.82	5900.15
Fe XIV	274.20	6.30	24752.87	2515.86	4497.33
Fe XV	284.16	6.30	507595.53	20607.53	35546.88

corresponding DN values to be estimated. Results for a selection of these lines are given in Tables 11 (quiet Sun), 12 (active region) and 13 (flares). For the flare line table, an extended version of the CHIANTI V4 differential emission measure was used. Earlier calculations used a version that cut off at log $T = 7.4$ K and thus did not account properly for Fe XXIV line fluxes. More complete tabulations are available on the MSSL EIS Website in the EIS planning guide at the URL given below.[1] At the EIS spectral resolution, the registered photons are actually spread out over three or more pixels. The complete tables include those

---

[1] http://www.mssl.ucl.ac.uk/www_solar/solarB/espg.html.

lines with detected photon numbers greater than 0.5 (quiet), 6.0 (active region) and 1000.0 (flare) photons. These tables allow estimates of the time needed to detect a given number of photons in a spectrum line. CHIANTI intensities have been calculated for optically thin emission lines. However, in a paper considering solar minimum EUV irradiance (Warren, 2005), it was noted based on Skylab observational values that when allowance is made for optical depth effects, the He II line intensities should be increased by a factor two over the CHIANTI values. This has been done for the entries in Tables 11 and 12.

8. Conclusions

The EIS is designed to study the high temperature plasma in the Sun's corona and upper transition region in the temperature range from below 1 MK to 20 MK and above. It has 2 arc sec spatial resolution and a plasma velocity measurement capability of better than ± 5 km s^{-1} for integration times of $10 - 100$ s for active region emission lines. Line profile studies will allow non-thermal effects or turbulent conditions in the plasma to be recognized. It will in addition measure plasma temperature and density and allow the construction of differential emission measure functions for the broad temperature range mentioned above. Measurements of element and ion abundances, particularly in outflowing plasma, will also be made. EIS achieves the above capabilities through the use of matched optimized multi-layer coatings on the primary mirror and on a focusing toroidal diffraction grating together with photon detection by back thinned and illuminated CCDs of high quantum efficiency in the 170 to 290 Å wavelength range. The ten times greater effective area than that of previous instruments in this spectral range, coupled with a higher data rate than was available from the SOHO spectrometers, will allow higher cadence studies of transient phenomena to be undertaken.

Together with XRT, EIS will benefit from the enormous capability of the SOT with its ability to measure photospheric velocity and vector magnetic fields at an angular resolution of ≈ 0.25 arc sec. Activity at the solar surface (photosphere) along with sub-surface activity in the convection zone, controls the upper atmosphere (transition region/corona) through the dynamic behavior of the photospheric plasma and the emergence of magnetic field from below the Sun's surface. Field lines project into the high atmosphere where they control the existence of the hot ($T > 1$ MK) coronal plasma and are responsible for the violent transient phenomena, $e.g.$, solar flares and CMEs, that have important effects on the near-Earth environment. The Hinode instruments are geared towards understanding the magnetic connection between the photosphere and underlying convection zone, and the corona, with particular reference to the phenomena of solar activity: structures, dynamics, plasma heating and transient events, to the evolution of the quiet sun network and intra-network regions and to the transfer of energy into the solar atmosphere. EIS in particular will focus on the dynamic and thermal response of the corona to the changing magnetic and velocity fields of the photospheric and sub-photospheric layers of the Sun.

The small fields of view of the SOT and EIS instruments will require careful joint observation planning that will be conducted on a monthly basis. Meetings will focus on (i) strategic planning for observations in the next three-month period and (ii) more detailed plans for the following month, subject to modification in response to developing solar conditions. Proposals for the use of EIS to make joint observations as part of the Hinode payload should be addressed to eis_obs@mssl.ucl.ac.uk.

The advanced features of the EIS instrument will enable it to play a highly significant part in the Hinode mission alongside the unique Solar Optical Telescope with its velocity and magnetic field observing capability together with the full-Sun context view provided at high time cadence by the high resolution X-ray telescope.

Acknowledgements Hinode is a Japanese mission constructed and launched by JAXA/ISAS, collaborating with NAOJ as a domestic partner, NASA (USA) and PPARC (UK) as international partners. Scientific operation of the Hinode mission is conducted by the Hinode science team organized at ISAS/JAXA. This team mainly consists of scientists from institutes in the partner countries. Support for the EIS construction and post-launch operation program is provided by PPARC (UK), NASA (USA), the Ministry of Education and Culture (Japan), the European Space Agency and the Norwegian Space Centre.

We would like to thank Sue Horne and Rosemary Young for their management of the PPARC funding. Among the MSSL staff, the contributions of Alec McCalden and Chris McFee are gratefully acknowledged. Work at the Naval Research Laboratory and Goddard Space Flight Center was supported by the NASA Marshall Space Flight Center (MSFC). In particular, we would like to thank MSFC's Hinode Program Manager, Larry Hill, for his support and guidance during the development, testing and integration of the EIS instrument. The authors wish to acknowledge the staff of JAXA/ISAS and NAOJ, along with the engineers of the companies involved in this project. Among them are K. Minesugi for mechanical design; A. Onishi and K. Hiraide for thermal design; M. Noguchi and M. Nakagiri for EIS testing at ISAS. We are particularly grateful to Bo Andersen of the Norwegian Space Centre for his efforts both on behalf of EIS and of the Hinode mission. JLC thanks the Leverhulme Foundation for the award of an Emeritus Fellowship. Finally we would like to acknowledge the substantial effort by the teams at MSSL, NRL, RAL, Birmingham, NAOJ, ISAS/JAXA and the University of Oslo who made the EIS instrument a reality.

References

Antiochos, S., DeVore, C.R., Klimchuk, J.: 1999, *Astrophys. J.* **510**, 485.

Beutler, H.G.: 1945, *J. Opt. Soc. Am.* **35**, 311.

Born, M., Wolf, E.: 1964, *Principles of Optics*, 2nd edn., MacMillan, New York, p. 469.

Canali, C., Martini, M., Ottavani, G., Alberigi Quaranta, A.: 1972, *IEEE Trans. Nucl. Sci. NS-19*.

Culhane, J.L., *et al.*: 1991, *Solar Phys.* **136**, 89.

Danzmann, K., Günther, M., Fischer, J., Kock, M., Kühne, M.: 1988, *Appl. Opt.* **27**, 4947.

Dere, K.P., Landi, E., Mason, H.E., Monsignori Fossi, B.C., Young, P.R.: 1997, *Astron. Astrophys. Suppl.* **125**, 149.

Engström, L.: 1998, GFit Version 10.0. *Lund Reports on Atomic Physics*, LRAP-232, LTH, Lund. http://kurslab-atom.fysik.lth.se/Lars/Gfit/html/index.html.

Finley, D.S., Bowyer, C.S., Paresce, F., Malina, R.F.: 1979, *Appl. Opt.* **18**, 649.

Haber, H.J.: 1950, *J. Opt. Soc. Am.* **40**, 153.

Hara, H.: 2003, *FM XRT-S Calibration*, National Astronomical Observatory of Japan.

Harra, L.K., Sterling, A.: 2003, *Astrophys. J.* **587**, 429.

Harra, L.K., Gallagher, P.T., Phillips, K.J.H.: 2000, *Astron. Astrophys.* **362**, 371.

Harra, L.K., Mandrini, C.H., Matthews, S.A.: 2004, *Solar Phys.* **223**, 57.

Harrison, R.A.: 1997, *Solar Phys.* **175**, 467.

Harrison, R.A., *et al.*: 1995, *Solar Phys.* **162**, 233.

Hollandt, J.: 1994, Strahlungsnormale für die solare Spektroradiometrie im Vakuum-UV, Ph.D. dissertation, Technischen Universität, Berlin.

Hollandt, J., Kühne, M., Huber, M.C.E., Wende, B.: 2002, In: Pauluhn, A., Huber, M.C.E., von Steiger, R. (eds.) *The Radiometric Calibration of SOHO, International Space Science Institute Report SR-002*, ESA, Nordwijk, p. 51.

Innes, D.E., Inhester, B., Axford, W.I., Willhelm, K.: 1997, *Nature* **386**, 811.

Kelly, R.L.: 1987, *J. Chem. Phys. Ref. Data* **16** (Suppl. No. 1).

Klimchuk, J.A.: 2006, *Solar Phys.* **234**, 41.

Korendyke, C.M., *et al.*: 2006, *Appl. Opt.* **45**, 8674.

Kowalski, M.P., *et al.*: 1999, *Appl. Opt.* **38**, 6487.

Kramida, A.E., Brown, C.M., Feldman, U., Reader, J.: 2006, *Phys. Scr.* **74**, 156.

Lang, J., Kent, B.J., Seely, J.F.: 2002, In: Pauluhn, A., Huber, M.C.E., von Steiger, R. (eds.) *The Radiometric Calibration of SOHO, International Space Science Institute Report SR-002*, ESA, Nordwijk, p. 337.

Lang, J., *et al.*: 2000, *J. Opt. A: Pure Appl. Opt.* **2**, 88.

Lang, J., *et al.*: 2006, *Appl. Opt.* **45**, 8689.

Madjarska, M.S., Doyle, J.G., van Driel-Gesztelyi, L.: 2004, *Astrophys. J.* **603**, L57.

Masuda, S., Kosugi, T., Hara, H., Tsuneta, S., Ogawara, Y.: 1994, *Nature* **371**, 495.

Nakariakov, V.M., Ofman, L.: 2001, *Astron. Astrophys.* **372**, L53.

NIST Atomic Spectral Data Base: 2005, http://physics.nist.gov/cgi-bin/asd.

Pevtsov, A.A.: 2000, *Astrophys. J.* **531**, 553.

Powell, F.: 1992, In: Bennett, H.E., Chase, L.L., Guenther, A.H., Newman, B., Soileau, M.J. (eds.) *Laser-Induced Damage in Optical Materials*, *Proc. SPIE* **1848**, 503.

Sakao, T., *et al.*: 2004, In: Mather, J.C. (ed.) *Optical, Infrared, and Millimeter Space Telescopes*, *Proc. SPIE* **5487**, 1189.

Samson, J.A.R.: 1967, *Vacuum Ultraviolet Spectroscopy Techniques*, Wiley, New York, p. 315.

Scholze, F., Henneken, H., Kuschnerus, P., Rabus, H., Richter, M., Ulm, G.: 2000, *Nucl. Instrum. Methods* **39**, 208.

Schumacher, R.J., Hunter, W.R.: 1977, *Appl. Opt.* **16**, 904.

Seely, J.F., Brown, C.M., Windt, D.L., Donguy, S., Kjornrattanawanich, B.: 2004, *Appl. Opt.* **43**, 1463.

Stern, R.A., *et al.*: 2004, In: Fineschi, S., Gummin, M.A. (eds.) *Telescopes and Instrumentation for Solar Astrophysics*, *Proc. SPIE* **5171**, 77.

Sze, S.M.: 1981, *Physics of Semiconductor Devices*, 2nd edn., Wiley, New York, p. 754.

Tsuneta, S.: 1995, *Publ. Astron. Soc. Japan* **47**, 691.

Tsuneta, S.: 1996, *Astrophys. J.* **456**, L63.

Ulm, G. Wende, B.: 1997, In: Haase, A., Landwehr, G., Umbach, E. (eds.) *Röntgen Centennial*, World Scientific, Singapore, p. 81.

Wang, T.J., Solanki, S.K., Curdt, W., Innes, D.E., Dammasch, I.E.: 2002, *Astrophys. J.* **574**, L101.

Wang, T.J., Solanki, S.K., Curdt, W., Innes, D.E., Dammasch, I.E., Kliem, B.: 2003, *Astron. Astrophys.* **402**, L17.

Warren, H.: 2005, *Astrophys. J. Suppl.* **157**, 147.

Warren, H., Doschek, G.A.: 2005, *Astrophys. J.* **618**, L157.

Williams, D.R., *et al.*: 2002, *Mon. Not. Roy. Astron. Soc.* **336**, 747.

Williams, D.R., Török, T., Démoulin, P., van Driel-Gesztelyi, L., Kliem, B.: 2005, *Astrophys. J.* **628**, L163.

Yokoyama, T., Akita, K., Morimoto, T., Inoue, K., Newmark, J.: 2001, *Astrophys. J.* **546**, L69.

Young, P.R., Del Zanna, G., Landi, E., Dere, K.P., Mason, H.E., Landini, M.: 2003, *Astrophys. J. Suppl.* **144**, 135.

The Solar Optical Telescope for the *Hinode* Mission: An Overview

S. Tsuneta · K. Ichimoto · Y. Katsukawa · S. Nagata · M. Otsubo · T. Shimizu ·
Y. Suematsu · M. Nakagiri · M. Noguchi · T. Tarbell · A. Title · R. Shine ·
W. Rosenberg · C. Hoffmann · B. Jurcevich · G. Kushner · M. Levay · B. Lites ·
D. Elmore · T. Matsushita · N. Kawaguchi · H. Saito · I. Mikami · L.D. Hill ·
J.K. Owens

Originally published in the journal Solar Physics, Volume 249, No 2.
DOI: 10.1007/s11207-008-9174-z © Springer Science+Business Media B.V. 2008

Abstract The Solar Optical Telescope (SOT) aboard the *Hinode* satellite (formerly called
Solar-B) consists of the Optical Telescope Assembly (OTA) and the Focal Plane Package
(FPP). The OTA is a 50-cm diffraction-limited Gregorian telescope, and the FPP includes the
narrowband filtergraph (NFI) and the broadband filtergraph (BFI), plus the Stokes Spectro-
Polarimeter (SP). The SOT provides unprecedented high-resolution photometric and vector magnetic images of the photosphere and chromosphere with a very stable point spread
function and is equipped with an image-stabilization system with performance better than
0.01 arcsec rms. Together with the other two instruments on *Hinode* (the X-Ray Telescope

M. Otsubo is a former NAOJ staff scientist.

S. Tsuneta (✉) · K. Ichimoto · Y. Katsukawa · S. Nagata · M. Otsubo · T. Shimizu · Y. Suematsu ·
M. Nakagiri · M. Noguchi
National Astronomical Observatory of Japan, Mitaka, Tokyo 181-8588, Japan
e-mail: saku.tsuneta@nao.ac.jp

T. Tarbell · A. Title · R. Shine · W. Rosenberg · C. Hoffmann · B. Jurcevich · G. Kushner · M. Levay
Lockheed Martin Solar and Astrophysics Laboratory, B/252, 3251 Hanover Street, Palo Alto,
CA 94304, USA

B. Lites · D. Elmore
High Altitude Observatory, NCAR, P.O. Box 3000, Boulder, CO 80307-3000, USA

T. Matsushita · N. Kawaguchi · H. Saito · I. Mikami
Communication Systems Center, Mitsubishi Electric Corp., Amagasaki, Hyogo 661-8661, Japan

L.D. Hill · J.K. Owens
Space Science Office, VP62, NASA Marshall Space Flight Center, Huntsville, AL 35812, USA

Present address:
S. Nagata
Kwasan and Hida Observatories, Kyoto University, Yamashina, Kyoto 607-8471, Japan

Present address:
T. Shimizu
Institute of Space and Astronautical Science, Japan Aerospace Exploration Agency, Sagamihara,
Kanagawa 229-8510, Japan

(XRT) and the EUV Imaging Spectrometer (EIS)), the SOT is poised to address many fundamental questions about solar magnetohydrodynamics. This paper provides an overview; the details of the instrument are presented in a series of companion papers.

Keywords Solar-B · Hinode · Sun: magnetic fields · Sun: photosphere · Sun: chromosphere · Sun: MHD

1. Introduction

The Sun has strong magnetic fields and emits intense X-rays from its outer atmosphere. Though observations with the *Yohkoh* satellite point to magnetic reconnection as a necessary ingredient for sporadic coronal heating on various scales from major flares to ubiquitous tiny bursts (Tsuneta, 1996; Yoshida and Tsuneta, 1996), the specific mechanisms of coronal and chromospheric heating remain essentially unknown. Recent progress from ground-based observations show that the solar magnetic field consists of an ensemble of fine-scale ($\approx 0.1'' - 0.2''$) magnetic fields in addition to sunspots and pores (Solanki, Inhester, and Schüssler, 2006). Detailed properties of solar magnetic fields are, however, still unknown owing to limitations of spatial resolution and accuracy of magnetic field measurements.

Solar magnetic fields are believed to arise as a result of a global dynamo operating at the base of the convection zone, and also possibly from a local dynamo process (Cattaneo, 1999). Ultimately, we need to improve our knowledge of the solar interior to fully understand the dynamo mechanisms. Even so, the emergence, dispersal, and decay of magnetic features at and above the solar photosphere provide an extremely valuable tool for exploring the mechanism of how magnetic flux is generated in the interior and is transported to the surface (Fisher *et al.*, 2000).

The main objective of the *Solar-B* (renamed *Hinode* after launch; Kosugi *et al.*, 2007; Figure 1) mission is to use a systems approach to understand the generation, transport, and ultimate dissipation of solar magnetic fields with a complex of three coordinated telescopes. For this purpose, *Hinode* carries the X-ray Telescope (XRT; Golub *et al.*, 2007; Kano *et al.*, 2007), the EUV Imaging Spectrometer (EIS; Culhane *et al.*, 2007), and the Solar Optical Telescope (SOT). The energy release and dissipation phase of the magnetic fields are observed with the XRT and EIS; the SOT performs high-resolution photometric and magnetic observations of the magnetic flux emergence and their subsequent evolution in the photosphere and chromosphere. The uniqueness of the *Hinode* mission is in using its coordinated and simultaneous observations of the photosphere, the chromosphere, the transition region, and the corona to understand how the changing photospheric and chromospheric magnetic fields results in the dynamic response of the coronal plasma.

In the early concept design phase of 1995 – 1996, the baseline configuration of the SOT was established to be a 50-cm diffraction-limited ($0.2'' - 0.3''$) telescope with both a filtergraph and a spectropolarimeter, by considering a balance between the scientific advantage over existing ground-based observations and technical constraints. The filtergraph was needed for high spatial and temporal resolution of the photometric and magnetic observations for both the photosphere and the chromosphere, whereas the spectropolarimeter was needed for precise observations of vector magnetic fields. In the course of the 10-year development, progress in high-resolution ground-based observations has been remarkable: The Swedish Solar Telescope (SST; *e.g.*, Scharmer *et al.*, 2002) delivered $\approx 0.1''$ photometric images and $\approx 0.2''$ longitudinal magnetograms. Spectropolarimetric observations with the German Vacuum Tower Telescope (VTT; *e.g.*, Bello Gonzalez *et al.*, 2005) and the Dunn

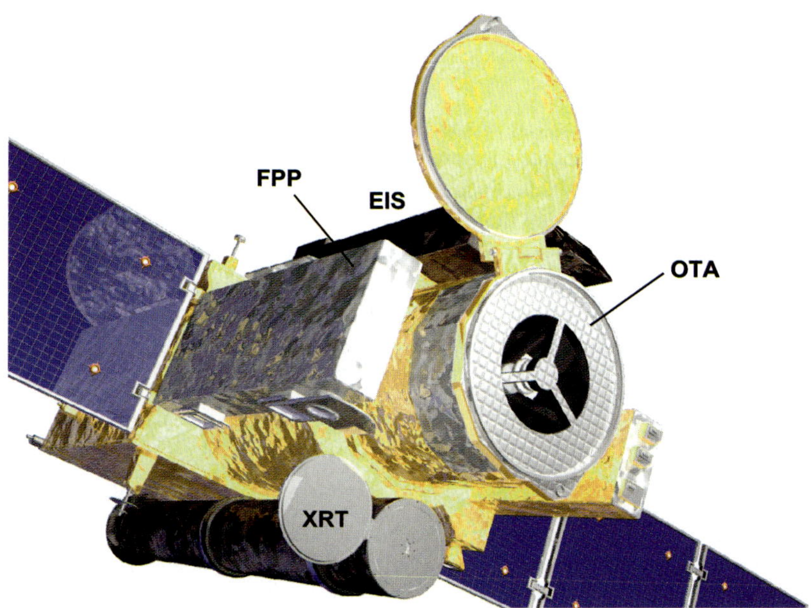

Figure 1 *Solar-B* (*Hinode*) outlook in orbit.

Solar Telescope (DST; *e.g.*, Lites, 1996) reached $\approx 0.4'' - 0.6''$ resolution. Indeed, the spatial resolution of the SST may be higher than that of SOT, and the spectropolarimetric resolution of the VTT and DST may be close to that of the SOT. Nevertheless, simultaneous photometric (imaging) and spectropolarimetric observations over extended periods of time ($>$ days) with a stable point spread function are critically important for almost all areas in solar studies. We stress the scientific importance of the uninterrupted observations (shown in movies) for understanding the ever-changing photospheric and chromospheric phenomena.

The 50-cm diameter SOT can obtain a continuous, seeing-free series of diffraction-limited images ($0.2 - 0.3$ arcsec) with fully calibrated high polarimetric sensitivity in the images with a broad spectral resolution (≈ 0.8 nm) in six wavelength bands at the highest resolution. The Narrowband Filter Imager (NFI) provides intensity, Doppler-shift, and vector-polarimetric imaging with moderate spectral resolution (≈ 10 pm) in nine spectral lines. When combined, the Broadband Filter Imager (BFI) and NFI observations cover the region from the low photosphere through the chromosphere. The Spectro-Polarimeter (SP) provides the line profiles in all Stokes parameters, with the high spectral resolution of 2.15 pm in two magnetically sensitive lines at 630.2 nm. For a typical exposure time, the sensitivity of the SP is $1 - 5$ G in the longitudinal direction and $30 - 50$ G in the transverse direction.

The time cadence ranges from tens of seconds for both photometric images and vector magnetograms in selected NFI lines to a few hours for wide-field scans with the SP. The maximum field of view for the NFI is $328'' \times 164''$ with a pixel size of $0.08''$, whereas that of the BFI is $218'' \times 109''$ with $0.053''$ per pixel, and the SP can view an area of $320'' \times 151''$ with a pixel size of $0.16''$ per pixel.

The Sun-synchronous orbit of *Hinode* makes possible uninterrupted observations for about eight months a year and is essential for providing a constant heating of the telescope, which is necessary for the optothermal stability of the telescope. The downlink of data in nearly every orbit through the ESA Svalbard station significantly contributes to the SOT science by allowing high cadence, a wide field of view, and a high-resolution observing program.

This paper provides an overview of the Solar Optical Telescope; accompanying papers describe the SOT key components in more detail: the Optical Telescope Assembly (Suematsu *et al.*, 2008), the Focal Plane Package (Tarbell *et al.*, 2008), the Image stabilization system (Shimizu *et al.*, 2008), and the instrument polarization calibration (Ichimoto *et al.*, 2008). In Sections 2 and 3, we overview the SOT science and technical system. The SOT consists of the Optical Telescope Assembly (OTA) and the Focal Plane Package (FPP), which are described in Section 4. Sections 5 and 7 present the observing modes, control, and data flows. Section 6 contains a brief description of the image stabilization system.

2. Science Overview

We discuss here some of the outstanding questions to be studied by the SOT and the *Hinode* observatory (Figure 2). Given the excellent quality of the data, we stress that in almost all research areas benefiting from the SOT, interaction with numerical simulations becomes critically important for the quiet Sun (Khomenko *et al.*, 2005), for emerging flux (Cheung, Schüssler, and Moreno Insertis, 2007), for chromospheric waves (Skartlien, Stein, and Nordlund, 2000), and for corona–chromospheric connections (Hansteen *et al.*, 2006; Abbett, 2006; Gudiksen and Nordlund, 2005).

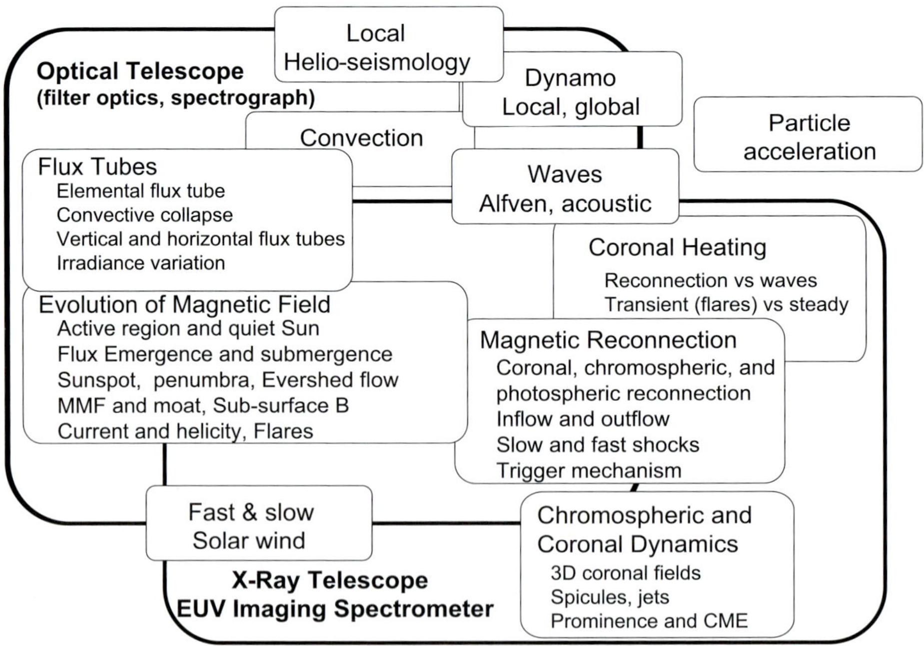

Figure 2 Scientific coverage of the SOT and the *Hinode* observatory.

2.1. Coronal Heating, Reconnection, and Waves: Synergy with XRT and EIS

The solar corona is believed to be heated by magnetic reconnection and/or dissipation of
MHD waves (Walsh and Ireland, 2003). Direct detection of the various modes of MHD
waves with the SP is within reach (Ulrich, 1996). Very high frequency MHD waves, if they
exist, may play an important role in the heating of active region corona. However, Parker
(1988) proposed that coronal heating is a consequence of reconnection of magnetic fields
that have become entangled as a result of motion of their photospheric footpoints. Whether
this is true or not should be answered observationally by Lagrangian tracking of individual
magnetic elements with the SOT. EIS may detect turbulence associated with jets from the
ubiquitous reconnection sites.

Yohkoh and *TRACE* (Handy *et al.*, 1999) images show spatially distinct hot and cool
quasi-steady loops, suggesting position-dependent heating rates. Using the HAO/NSO Advanced Stokes Polarimeter, Katsukawa and Tsuneta (2005) found a clear difference in the
magnetic filling factor, which is the areal fraction of magnetic atmosphere, at the footpoints
of hot and cool loops. The SOT allows one to better resolve the specific photospheric conditions (*i.e.*, flows and fields) resulting in hot and cold coronal structures, leading to a deeper
understanding of coronal heating.

2.2. Active Regions and Sunspots

The magnetic field in the photosphere is distributed in a very inhomogeneous way, with
sunspots and faculae being the centerpieces of active regions (Solanki, 2003; Weiss, 2006;
Ferriz-Mas and Steiner, 2008). There are a number of obvious questions to be pursued with
the SOT: How are the basic umbral and penumbral structures of sunspots formed and maintained? What drives the Evershed flow in the sunspot penumbral photosphere and the oppo-
sitely directed inverse Evershed flow in the penumbral chromosphere? How do they disinte-
grate and spread their magnetic fragments to the quiet Sun – possibly in a form of moving
magnetic features (MMF)? What is the relationship among umbral dots, light bridges, and
convection? High-resolution, precise, and continuous observations by the SOT are uniquely
contributing new information about many features, including sunspots, moat regions, umbral
dots, light bridges, and their subsurface structures.

An important product of SOT's Dopplergram capability is the three-dimensional maps of
subsurface flows and magnetic fields that can be obtained through the application of local
helioseismology. This extends our investigation to subsurface layers, where much of the
action is taking place (Sekii, 2004; Kosovichev, 2004).

2.3. Flux Tubes and Quiet-Sun Magnetic Fields

A ubiquitous form of magnetic fields at the photospheric level is that of small-scale, unipolar
vertical kilogauss fields, sometimes observed as bright points in the G band (Berger *et al.*,
2004). Convective collapse (Parker, 1978) may form the kilogauss-strength tubes and eventually form pores and sunspots from weaker emerging fields. However, we do not yet know
how they are created, how they evolve, and how they are destroyed. Supergranular diffusion
and poleward meridional flow was believed to transport fragmented magnetic fields away
from sunspots and active regions and to provide magnetic flux to the quiet Sun (Leighton,
1964). In addition to this, we now know that numerous bipolar ephemeral regions with life-
times of several hours (Harvey-Angle, 1993) and ubiquitous small-scale horizontal magnetic
fields (Lites *et al.*, 1996) with much smaller time scales emerge and submerge, and, as a result, magnetic fields in the quiet Sun are quickly replaced (Title, 2007). Stable long-term

observations with the SOT could clarify the demography of these magnetic elements with different origins.

2.4. Data-Driven Simulation of Coronal Dynamics

Accurate vector-magnetic images obtained with the SOT provide us with time-dependent boundary conditions for coronal magnetic fields. These images allow us to construct 3-D extrapolations of magnetic fields into the corona initially as a snapshot, and eventually a time-dependent evolution is constructed (Welsch *et al.*, 2007). One cautionary note is that there is a tradeoff between polarimetric accuracy and the time required to scan an active region: At least for small-scale flux elements the SP scan duration is usually larger than the time scale for change of a solar feature. If our extrapolations are successful in reproducing the magnetic field structure with electric current sheets in the corona, the stability analysis as well as the data-driven simulation of the solar coronal dynamics would make possible the forecasting of flares and CMEs in response to the evolution of surface magnetic fields.

2.5. Chromospheric Heating and Dynamics

The chromosphere is maintained by an energy flux ≈ 10 times greater than that required to maintain the corona (Withbroe and Noyes, 1977). The observational signature for either wave heating (and/or resultant shocks) or magnetic reconnection (or both) should be observed with the SOT by using both photospheric and chromospheric lines with co-temporal magnetograms (Ulmschneider and Musielak, 2003; Carlsson and Stein, 2004). The chromosphere is highly dynamic, showing ubiquitous jets such as spicules, which can supply mass to the corona and the solar wind. Coordinated SOT and EIS observations studying the thermal evolution of chromospheric ejections are important (Sterling, 2000; De Pontieu, 2004). Furthermore, the chromosphere is closer to the force-free corona, and the chromospheric magnetic fields obtained with the SOT potentially can give better boundary conditions for the coronal field extrapolation than those from the non-force-free photosphere (Metcalf *et al.*, 1995; Leka and Metcalf, 2003).

3. System Overview

The Solar Optical Telescope is detailed in a series of figures: Figure 3 shows a schematic diagram of the optical systems, Figure 4 shows the electrical configuration, and Figure 5 shows the optical schematic diagram. The OTA and the FPP (as are XRT and EIS) are mounted on the satellite optical bench (OBU; Figure 6), which is stable against the launch (mechanical) and orbital (thermal) environment. Figure 7 shows the optical interface between the OTA and the FPP. An accurate alignment of images from the three telescopes is crucial, and extensive testing to characterize the thermal deformation of the OBU was carried out to ensure the necessary stability. As a result, SOT images are accurately aligned with XRT and EIS images through the observatory-level alignment procedure, which employs successive ladders through nearby images (in terms of wavelength) taken with the different telescopes.

The OTA (Figure 8) consists of the primary mirror, secondary mirror, Heat Dump Mirror (HDM), Collimator Lens Unit (CLU), secondary Field Stop (2FS), Tip-tilt fold Mirror (CTM-TM), and the Polarization Modulator Unit (PMU). Located in the OTA, the PMU is controlled by the FPP through its critical timing with that of CCD exposures needed for polarization modulation. The OTA has two deployment doors for the heat dump window and

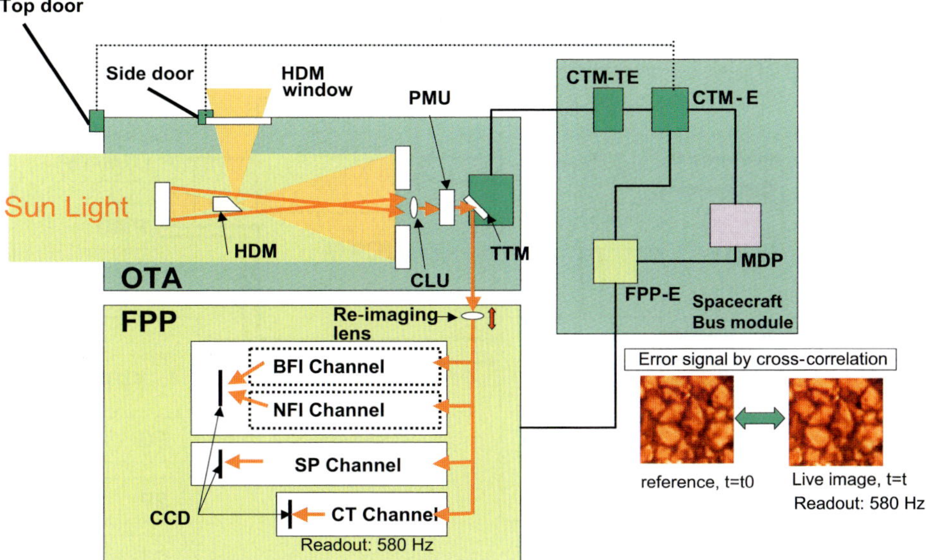

Figure 3 SOT system overview: OTA – Optical Telescope Assembly, FPP – Focal Plane Package, HDM – Heat Dump Mirror, CLU – Collimator Lens Unit, PMU – Polarization Modulation Unit, TTM (referred to as CTM-TM in the text) – Tip-tilt fold Mirror, BFI – Broadband Filter Imager, NFI – Narrowband Filter Imager, SP – Spectro Polarimeter, CT – Correlation Tracker, MDP – Mission Data Processor, FPP-E – main electronics box for the FPP, CTM-E – main electronics box for the OTA and the TTM, and CTM-TE – analog driver for the TTM.

the entrance aperture, which serves as the entrance pupil. The OTA main structure, to which are mounted these critical optical components, is a precision truss made of zero-expansion graphite-cyanate composite material.

The FPP (Figure 9) has a reimaging lens followed by a beam splitter. The effective combined focal length is 1550 cm (f/31), and the depth of focus in the FPP focal plane is about 400 μm. The focus is adjusted by moving the reimaging lens through commands from the ground. The reimaging lens has a stroke of ± 25 mm, which gives it a sufficient margin based on the focus budget breakdown table, which has numerous deterministic and statistical factors.

On the downstream side of the beam splitter are the broadband and narrowband filter channels sharing a common CCD camera, the spectropolarimeter, and the correlation tracker. A nonpolarizing beam splitter divides the light between the SP and the filtergraph, and then the polarizing beam splitter in the filter channel transmits the p-polarized light to the NFI and the s-polarized light to the BFI.

The FPP electrical box (FPP-E) has a computer for controlling the FPP and performs onboard data processing such as Stokes demodulation. The other electrical box (FPP-PWR) contains the power supply for the entire FPP subsystem. The Mission Data Processor (MDP) controls FPP observations. It follows the observation tables located in the MDP and processes science and housekeeping data from the FPP. The CTM-E box has another dedicated computer for servo control with piezo-driver electronics (CTM-TE box) for driving the tip-tilt mirror (CTM-TM) located inside the OTA. The FPP computer and the CTE-M computer directly communicate with each other (*i.e.*, handshake, without any involvement

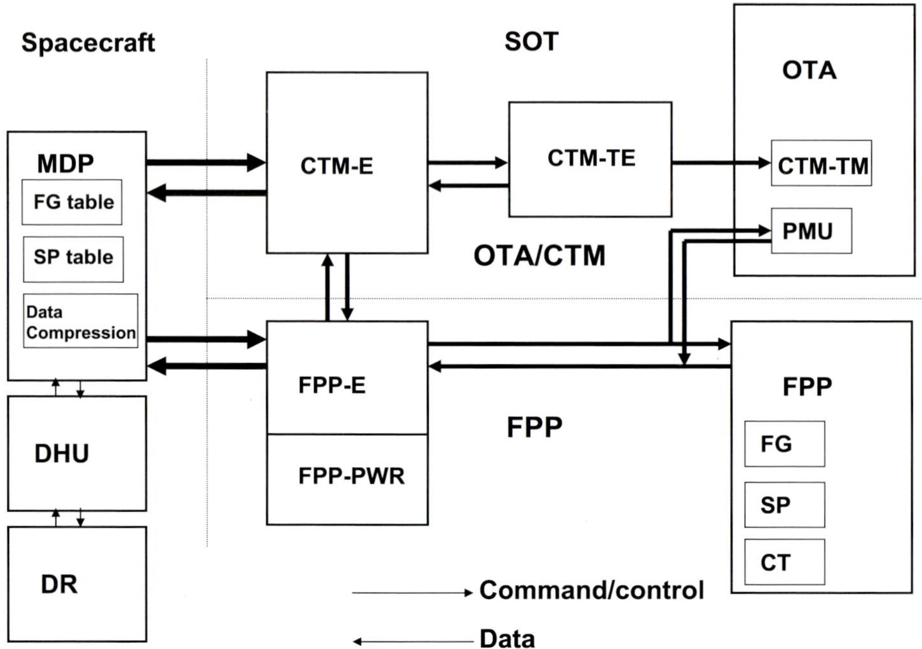

Figure 4 SOT subsystem instrument configuration: OTA – Optical Telescope Assembly, FPP – Focal Plane Package, PMU – Polarization Modulation Unit, CTM-TM – Tip-tilt fold Mirror, MDP – Mission Data Processor, FPP-E – main electronics box for the FPP, FPP-PWR – power supply for the FPP subsystem, CTM-E – main electronics box for the OTA and the TTM, CTM-TE – analog driver for the TTM, DHU – spacecraft Data Handling Unit, and DR – spacecraft central Data Recorder.

from the MDP) to close the control loop for image stabilization. All the commands from the ground to the FPP and the CTM-E go through the MDP.

The temperatures of the instruments directly affect instrumental safety and the optical performance for both the OTA and the FPP. Maintaining the instrument temperatures within the desired ranges is one of the critical functions of the system. There are numerous temperature sensors, some of which are fed to the servo controller. The OTA has operational heaters to maintain the temperatures of critical optical components and decontamination heaters to maintain the temperatures of critical optics higher than those of the telescope environment before opening the primary door. The CTM-E controls operational and decontamination heaters for the OTA. The FPP zone heaters maintain the entire FPP assembly at 20°C ± 1°C and also has decontamination heaters for CCD bakeout. The OTA and the FPP have survival heaters controlled by the spacecraft heater control electronics in case the primary power for the science instruments is cut off, thus ensuring spacecraft survival.

The observing tables are uploaded from the ground and provide extremely flexible observing sequences for the SOT and the XRT (for details on autonomous XRT observing control with the MDP, see Kano *et al.* (2007)). The MDP sends the SOT/FPP macro-commands, which contain all the parameters and instructions to perform the desired observations based on the uploaded tables. The housekeeping data and the image data with header information are separately sent to the MDP from the SOT. The image data are compressed, if instructed to do so, combined with the final header information, packetized, and sent to the spacecraft data recorder through the spacecraft central Data Handling Unit (DHU).

Optical layout of SOT

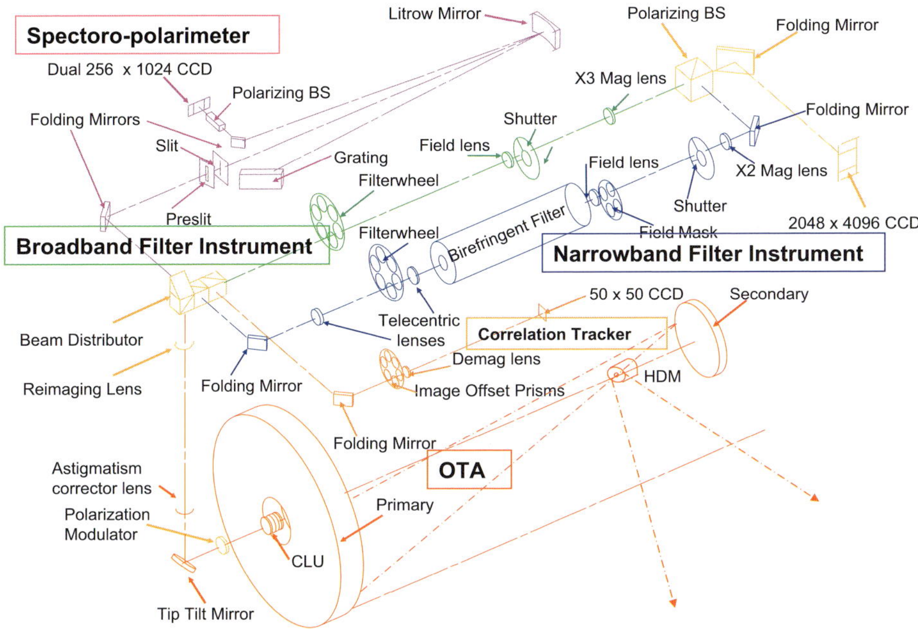

Figure 5   Optical layout of the SOT including the OTA and the FPP.

Figure 6   The OTA and the FPP mounted on the spacecraft optical bench (OBU). The cylindrical optical bench also carries the EIS and XRT instruments (not mounted in this photo) and is mounted on the spacecraft bus box. The FPP radiators are covered with red-colored protective covers.

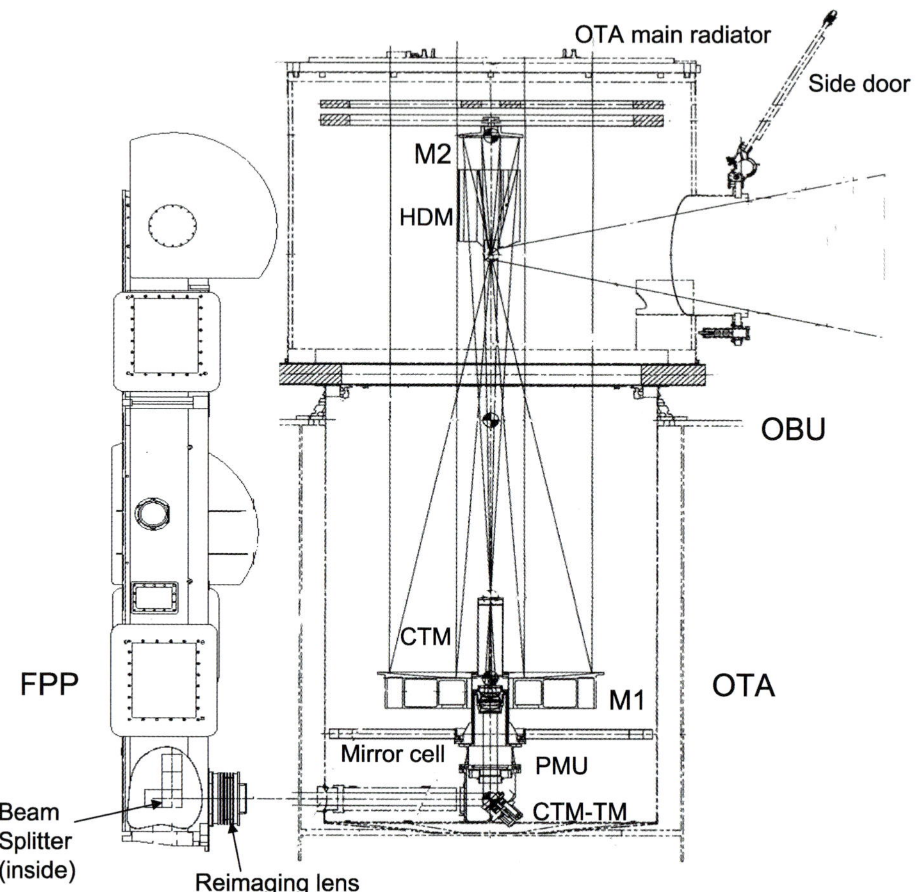

Figure 7 Optical interface between the OTA and the FPP. The OTA and the FPP are mounted on the common optical bench unit (OBU). Shown are M1 – OTA primary mirror, M2 – OTA secondary mirror, the HDM – Heat Dump Mirror, and the CLU – Collimator Lens Unit (taken from Suematsu *et al.*, 2008).

The MDP also stores the orbital elements of the spacecraft orbit and the information on the spacecraft pointing, which is used to calculate the Doppler shift of the solar spectral lines. These are sent to the SOT from the MDP, where they are used to compensate in a real-time manner the Doppler shift in the NFI observations, primarily resulting from satellite motion, with a tunable filter. Thus, the MDP plays various crucial roles in obtaining smooth and stable SOT observations.

4. Optical Telescope Assembly and Focal Plane Package

4.1. OTA Optics

The OTA is the diffraction-limited aplanatic Gregorian telescope with a 50-cm-aperture primary mirror (Suematsu *et al.*, 2008). Table 1 summarizes the main characteristics of the

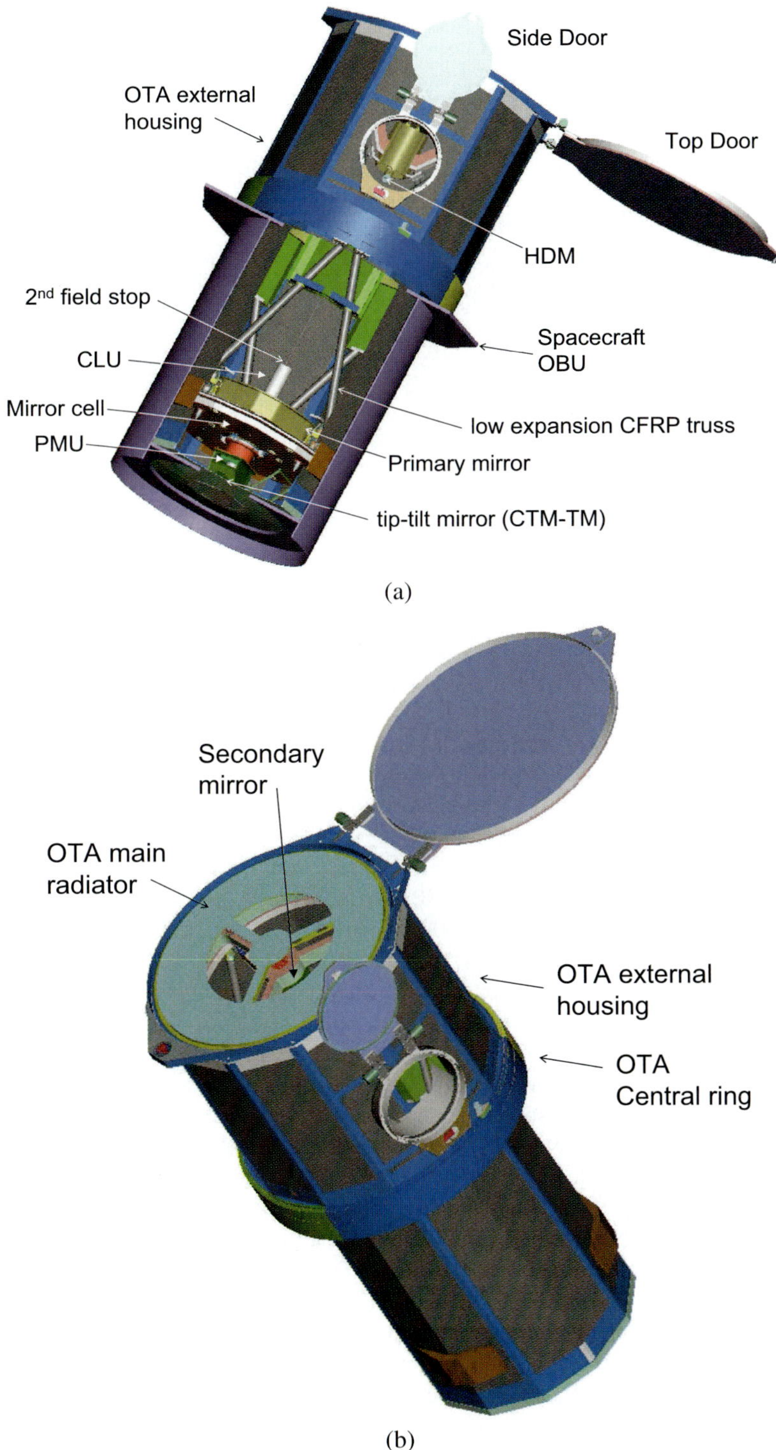

Figure 8 SOT Optical Telescope Assembly.

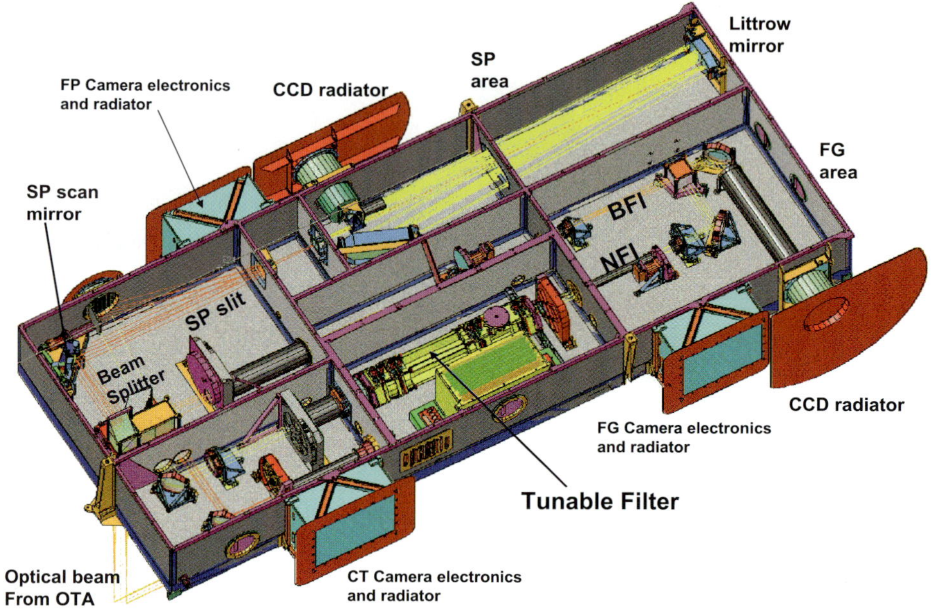

Figure 9   SOT Focal Plane Package (see Tarbell *et al.*, 2008).

OTA. The spatial resolution is specified in terms of the Strehl ratio: The Strehl ratios of the OTA and the FPP should be individually better than 0.9 at 500 nm, and the combined Strehl is higher than 0.8. Following a conventional definition of the diffraction limit (*e.g.*, Maréchal criterion), we set these numbers as a goal. The optical tests simulating the in-orbit condition of the OTA on the ground (Section 4.3) demonstrated that the OTA had a Strehl ratio better than 0.9 at 500 nm, and the measured FPP Strehl averaged over the field of view is very close to or exceeds 0.9 (Tarbell *et al.*, 2008). The post-launch performance appears to meet the goal and is described in Suematsu *et al.* (2008).

The primary and secondary mirrors are made from ULE. The lightweight (14 kg) primary mirror is supported by an elaborate kinematic mount system to fully meet the stringent requirements on the surface deformation over a wide range of temperatures. At the same time, the fragile primary mirror with its mount system had to survive the severe launch conditions (vibration and acoustic and shock loads) of the ISAS/JAXA M-V solid-booster launch vehicle. The secondary mirror is supported by a fixed invar/titanium mount. Both mirrors are coated with protected silver.

The distance between the primary and secondary mirrors was set at 1.5 m, by considering the amount of space in the crowded spacecraft, the optomechanical tolerance, and the manufacturability of the low-f-number primary mirror. The aluminum HDM at the primary focus reflects the unused solar light (heat) outside of the 400″-diameter field of view (FOV) into space through its side window. Since intensities at the HDM are 1500 times solar values, special development and testing efforts were required for its enhanced silver coating. The moderate temperature ($20°C - 40°C$) of the HDM is achieved through a high reflectivity and an innovative radiation-cooling design of the mirror.

The conical field stop is located at the secondary focus to limit the field of view to 361 × 197 arcsec. (Note that the widest observing field of view is 328 × 164 arcsec.) The design

Table 1 Optical Telescope Assembly overview.

Telescope	
Optics type	Aplanatic Gregorian with heat dump mirror
Primary mirror	50-cm aperture, lightweight ULE
Primary-to-secondary mirror length	1.5 m
Central obscuration ratio	0.344 in radius
Effective F ratio	9.055 at secondary focus
Coating	Protected silver coating for primary and secondary mirrors
Heat Dump Mirror (HDM)	
Mirror	Aluminum with enhanced silver coating Provides FOV of 400$''$
Collimator Lens Unit (CLU)	
Purpose	Create exit pupil, deliver parallel light to FPP
Focal length	37 cm
Wavelength	380 – 700 nm, achromatic for nominal temperature range
Exit pupil size	3 cm, collimated in air
Polarization Modulator Unit (PMU)	
Location	Located near exit pupil of OTA
Rotation speed	Continuous, 1.6 s rotation^{-1}
Retardation	1.35 waves for 630.2 nm
Coating	Enhanced silver coating
Tip-tilt mirror for image stabilizer (CTM)	
	See Table 4

ensures that the light discarded by the secondary field stop will be reflected back via the same route through the secondary and primary mirrors to space.

Behind the secondary focus of the OTA are the CLU, the PMU, and the CTM-TM. The CLU has a short focal length of 37 cm to deliver parallel light to the FPP and to create the exit pupil (*i.e.*, the image of the entrance pupil) in the vicinity of the PMU and the CTM-TM. This location of the exit pupil was quite fortunate for the SOT program: In the system-level optical test, we discovered an unacceptable degree of astigmatism – probably caused by the primary mirror. Very late in the program, we decided to add corrective optics (a single cylindrical lens) at the exit pupil to completely remove the astigmatism, which minimized the hardware change.

The CLU consists of six lenses with an IR-rejection filter at its entrance and is aberration-free (achromatic) and practically instrument-polarization free for the entire range of observing wavelengths (380 – 700 nm). The first two lenses, which are radiation-robust fused-silica, are used to protect four inner lenses that are more susceptible to radiation, and the spacecraft bus module behind the CLU serves as a backside radiation shield. The CMT-TM has an enhanced silver coating.

The extremely severe positional tolerance of the secondary mirror, with respect to the primary mirror, had been a concern in the design phase. However, the possibility of introducing an adjustment mechanism for the secondary mirror was not a viable choice owing to the limited resources in the OTA development program. During the course of the development, the choice to use a 50-cm mirror turned out to be close to the limit in terms of science, technology, cost, and the stringent constraint of construction time. Also, the 50-cm primary mirror is an intended scientific compromise between the requirements for high spatial resolution and a field of view large enough to cover a typical active region.

4.2. Polarization Modulation

The PMU, located near the exit pupil, is a continuously rotating waveplate with revolution period T of 1.6 s to provide polarization modulation. The temperature dependence of the PMU retardation is minimized by utilizing two crystals of compensating thermal coefficients of birefringence: quartz and sapphire. The retardation is wavelength-dependent but is optimized for the 630.2-nm (with a retardation of 1.35 waves) and 517.2-nm (1.85 waves) observations in the sense that Stokes vectors Q, U, and V have an equally high modulation efficiency of approximately 0.5. The modulation efficiency at other wavelengths is unbalanced among Stokes Q, U, and V.

The polarization states are represented by the Stokes vectors (I, Q, U, V). The linear polarization Q and U and the circular polarization V are converted into sinusoidal variations of intensity by the polarizing beam splitters in the FPP. Stokes Q, U, and V are encoded as harmonic variations of intensity at periods proportional to $T/4$, $T/4$, and $T/2$, respectively. The signal vector Q differs in phase from the signal U by 22.5 degrees (relative to the rotational phase of the waveplate). Demodulation of this signal is done by sampling the intensity 16 times per revolution of the PMU waveplate. I, Q, U, and V spectra are then obtained by either adding or subtracting each sample into the four memories corresponding to the four Stokes states in the FPP.

The rotating PMU is completely invisible for nonmagnetic photometric observations, and any movement in the image caused by its minimum residual wedge is removed by the image-stabilization system. All optical elements prior to the PMU are rotationally symmetric about the optical axis (except for the secondary mirror supports) to minimize instrumental polarization. Note that the folding tip-tilt mirror follows the PMU.

The FPP has a reimaging lens after receiving the parallel light from the OTA, and the OTA can therefore be regarded as the pupil reducer. In other words, the optical interface between the OTA and the FPP is intended to be afocal, considerably relaxing the positional tolerance of the FPP with respect to the OTA. This allows the OTA and the FPP to be separately mounted on the OBU as separate independent instruments without any precision requirement (Figure 6).

4.3. Optical Testing

In addition to the usual tests for space flight hardware such as vibration and thermal-vacuum tests, we performed a number of unique tests for the OTA and the FPP first separately and later jointly. Tremendous efforts were expended to plan, develop, and implement these tests, and some major problems – including the discovery of astigmatism – were found as a result of these tests. All of the problems found in the tests were completely analyzed, and corrective actions, sometimes requiring a hardware fix, were thoroughly taken, with the problems declared to be closed only after retesting demonstrated the desired performance. A lesson we learned is that extensive and complete testing is essential to success of an advanced space-optics instrumentation mission.

Wavefront-error (WFE) measurements comprise the fundamental test done in the laboratory environment and in the thermal vacuum chamber. A large rotatable precision optical flat was located in front of the OTA entrance aperture to measure the telescope WFE (using a double pass) under autocollimation. The interferometer is located at the position of the FPP to measure the OTA WFE. The WFE maps reveal a large triangle astigmatism mainly from gravity deformation of the primary mirror, but this was cancelled out by adding another WFE map taken with the OTA and the optical flat in the upside-down configuration. This methodology provides us with a means to measure the WFE in a zero-gravity situation. This test was repeated many times during the alignment of critical optical elements, as well as after mechanical tests such as vibration and shock tests.

The Sun test is a unique start-to-finish observation of real sunlight, which was introduced to the clean room through the heliostat by combining the OTA and the FPP with the flight electronics (Figure 6). A polarization calibration to obtain the Muller matrix of the SOT as a whole (OTA plus FPP) was carried out extensively by using the linear and circular polarizers at the entrance aperture of the OTA telescope (Ichimoto *et al.*, 2008).

The optothermal test was done to measure the OTA WFE with the large optical flat in the thermal vacuum chamber. Only the interferometer was located outside the chamber in this configuration. A special shroud was prepared to simulate the expected high temperature gradient along the optical axis while in orbit, and with this setup we were able to verify the OTA optical performance in an environment close to that for the actual observations. Other system-level (post-assembly level) tests for the OTA included a temperature cycle test, a vignetting test, a scattered light measurement, a focus test, and a throughput measurement.

After delivery of the SOT to the spacecraft systems, the SOT (as well as all the other instruments) was integrated to the *Solar-B* spacecraft, and the final series of tests – including vibration and shock tests – continued at ISAS for about one year, at which time instrument builders usually no longer have any access to the instrument to confirm the critical optical performance. However, the SOT is equipped with an optical maintenance port, which is a small hole on the OBU located around the OTA-FPP optical interface (Figure 7). Even after the full installation of the SOT to the spacecraft, we were able to measure the WFE of the OTA with the optical flat located in front of the OTA aperture and with the interferometer at the port (where we have a special optical GSE to introduce light to the OTA-FPP optical interface). The maintenance port was also used to check that the FPP CCD functioned and to confirm the FPP internal and external alignments. Via the optical maintenance port, we repeatedly had the opportunity to check the optical health of the OTA and the FPP, especially after harsh environment tests, up to delivery of the *Solar-B* spacecraft to the launch site. This was a tremendous help and bolstered our confidence in the in-orbit performance of all the systems.

4.4. Structural and Thermal Properties

4.4.1. OTA

The OTA has a precision truss structure consisting of graphite-cyanate composite-material pipes and honeycomb panels, which were developed especially for the OTA. They have a very low coefficient of thermal expansion (0.05 ppm per 1°C). The main structure is unique in the sense that all the building blocks are essentially connected with adhesive (not with bolts and pins) to have the required dimensional stability against temperature change and severe mechanical environments (vibration and shock). The OTA had to be lightweight because of the stringent weight budget situation of the spacecraft, but it had to withstand the violent vibration, acoustic, and shock loads imposed by the ISAS/JAXA M-5 launch booster. The total weight of the OTA is about 103 kg, and the FPP is about 46 kg (without including the separately located electrical boxes).

The lightweight ULE primary mirror is mounted on the mirror cell (honeycomb plate) through the kinematic mount mechanism, and the CLU, PMU, CMT-TM are also tightly mounted on the mirror cell. The secondary mirror and HDM are mounted on the spider structure located on the other end of the telescope. The kinematic interface to the cylindrical OBU is through the stiff central ring plate located near the midpoint of the telescope's structure. The main structure of the telescope is surrounded by the telescope external housing, which is mounted only on the central ring. The external housing has the entrance aperture (entrance pupil) and heat dump window. Those two apertures have deployment doors to maintain an ultra high level of cleanliness of the telescope during testing and launch and to prevent an invasion of sunlight into the telescope during the initial outgassing period in orbit.

Although sunlight outside the field of view is reflected into space by the HDM, optical elements such as the primary mirror illuminated by intense sunlight inevitably absorb some fraction of the energy. The solar absorptance is about 6.5% for the primary mirror. All of the optical elements are radiatively coupled to the telescope structure, and the HDM has special fins to dump the absorbed heat. The main internal heat source is the primary mirror located aft of the telescope and the HDM. The heat absorbed by these optical elements is eventually dumped through the large Sun-facing radiator (OSR) at the entrance aperture on the Sun side, since the backside of the OTA is occupied by the spacecraft bus module, and the OTA is thermally decoupled from the spacecraft. A heat pipe is not used in the telescope system, and the radiation coupling is the primary heat transfer path.

A high temperature gradient along the Z-axis (from 30°C at the primary mirror to below 0°C at the secondary mirror) is needed to transport the heat from the aft to the forward section. The OTA's temperature was determined by designed balance between the solar heat input and the heat dump efficiency from the inside of the OTA, which is regarded as a thermal cavity. Extensive efforts were made to experimentally verify this unique thermal design concept by utilizing two large-scale spacecraft-level thermal vacuum tests (Figure 10). In the thermal balance test of the proto-model OTA and the spacecraft, solar heat input to the individual optical components was simulated accurately by nonflight heaters attached to the optical elements.

There are three different heater systems to maintain the OTA temperatures. Operational heaters maintain the temperature of optical components with special temperature requirements. For example, the temperature of the CLU has to be kept above 25°C to avoid instrument polarization.

4.4.2. FPP

The FFP structure consists of an aluminum honeycomb optical bench with side panels and a cover plate. Note that because of the large depth of focus, using aluminum with large CTE does not pose a problem. The FPP box is mounted on the OBU by a spacecraft-provided kinematic mount. The FPP is thermally isolated from the OBU. Each CCD detector is cooled by its own dedicated radiator. The thermal and structural design of the FPP is described in detail by Tarbell *et al.* (2008).

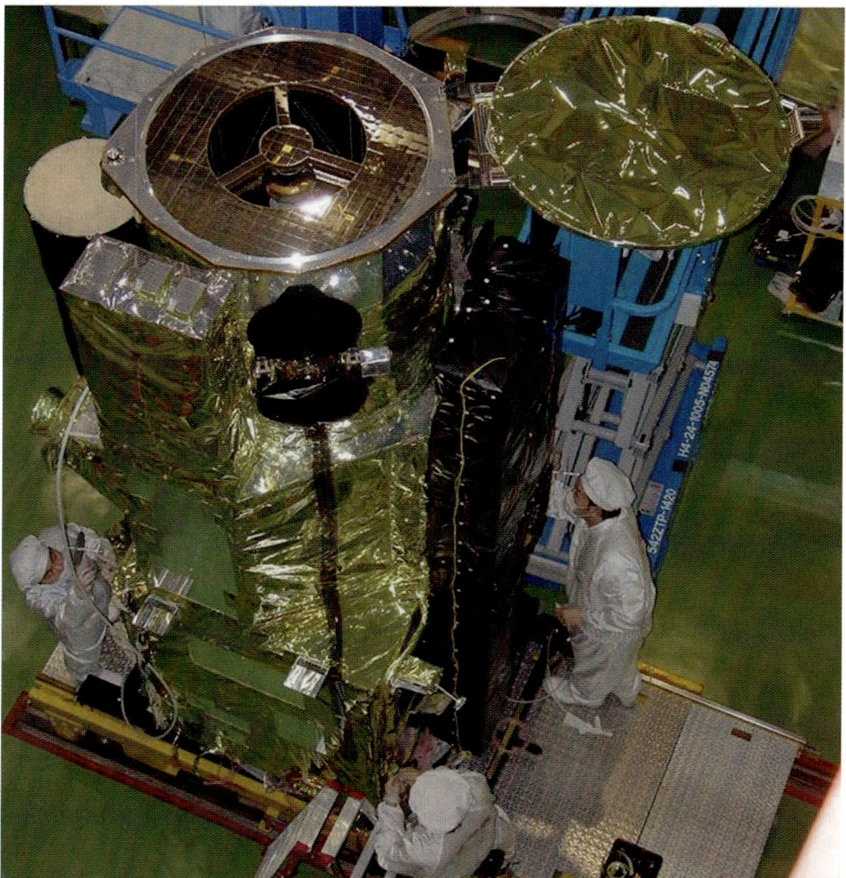

Figure 10 The SOT and the *Hinode* satellite prior to the spacecraft-level thermal vacuum test.

4.5. Contamination Control

A stringent contamination control program has been implemented from the early design phase through testing and launch to avoid any increase of heat input from the slight darkening of the optical surfaces when contaminated with organic materials. The temperature increase of the primary mirror would result in deformation of the mirror figure owing to a difference in CTE between the ULE glass and attached super-invar pads. Therefore, all the flight components are thoroughly baked out and their final outgas rates are quantitatively monitored with the Thermoelectric Quartz Crystal Microbalance. The OTA orbital lifetime (in terms of contamination degradation) is predicted by means of the OTA mathematical contamination model by using the measured outgas data.

The primary and secondary mirrors and the HDM have dedicated decontamination heaters that maintain the temperatures of the critical optical components at least 10°C higher than their surroundings during the high-outgas phase after launch and during the backfill period of the thermal vacuum test. Only the side door was quickly opened to vent the gas from the telescope after launch. The mathematical contamination model was used extensively to predict the outgassing period, after which the telescope main door was opened to introduce sunlight and enable the start of observations.

Note that the FPP does not necessarily need to have such a stringent plan for contamination control because hazardous solar UV light is essentially absorbed by the OTA, and the FPP has a closed structure without any exposed aperture.

5. SOT Observing Modes

With the CLU and the tip-tilt fold mirror, the OTA delivers a pointing-stabilized parallel beam to the FPP. The FPP is configured with the reimaging lens followed by the beam splitter for the filtergraph, the spectropolarimeter, and the correlation tracker channels. The FPP performs both filter (FG) and spectral (SP) observations at high polarimetric precision, and both types of observation may be performed simultaneously, yet independently, in response to the macro-commands from the MDP.

In filter observation, a 4k × 2k CCD camera is shared by the BFI and the NFI, which are selected by a common mechanical shutter. The SP and Correlation Tracker (CT) have their own CCD detectors. The NFI uses a tunable birefringent filter (Lyot filter) to record filtergrams, Dopplergrams, and longitudinal and vector magnetograms across the spectral range from 517.0 to 657.0 nm, including several spectral lines: Mg I b (517.3 nm; chromospheric Dopplergram and magnetogram), Fe I (525.0, 524.7, and 525.0 nm), Fe I (557.6 nm), Na I (589.6 nm; chromospheric Dopplergram and magnetogram), Fe I (630.3 and 630.2 nm), and H I (656.32 nm; chromospheric structure). The BFI has interference filters to image the photosphere (CN 388.3 nm and CH 430.5 nm) and low chromosphere (Ca II H 396.9 nm) and to make blue (450.5 nm), green (555.1 nm), and red (668.4 nm) continuum measurements for irradiance studies.

The SP is an off-axis Littrow–Echelle spectrograph that records dual-line (Fe I 630.25 nm and Fe I 630.15 nm) dual-beam (with the polarization beam splitter, which is a polarization analyzer, in front of SP CCD) Stokes spectra for high-precision Stokes polarimetry.

The time sequencing of the science data acquisition by the SOT is controlled according to observation tables (one for the FG and the other for the SP) on the MDP, as will be described in Section 7.

5.1. Filter Observations

The BFI (Table 2) produces photometric images with broad spectral resolution in six bands (CN band, Ca II H line, G band, and three continuum bands) at the highest spatial resolution available from the SOT (0.0541 arcsec pixel^{-1} sampling) and at a rapid cadence (< 10 s) over a 218×109 arcsec FOV. Exposure times are typically $0.03-0.8$ s, but longer exposures are possible, if desired. The BFI allows accurate measurements of horizontal flows and temperature in the photosphere, and measurements in the ultraviolet bands will permit identification of sites of strong magnetic field. The BFI observes the Ca II H line around line center. These BFI filters have FWMH bandwidth of $0.3-0.7$ nm and obtain images not subject to Doppler motion.

The NFI (Table 2) provides intensity, Doppler, and full Stokes polarimetric imaging at high spatial resolution (0.08 arcsec pixel^{-1} – somewhat coarser sampling than the BFI) in any one of 10 spectral lines (including Fe lines with a range of sensitivity to the Zeeman effect, Mg I b, Na D lines, and Hα) over the full field of view (328×164 arcsec). The

Table 2 SOT/FPP filter observations.

Broadband Filter Imager (BFI)

Field of view	218×109 arcsec (full FOV)
CCD	$4k \times 2k$ pixels (full FOV), shared with NFI
Exposure time	$0.03 - 0.8$ s (typical)
Spatial sampling	0.0541 arcsec pixel^{-1} (full resolution)

Spectral coverage

Center (nm)	Width (nm)	Line of interest	Purpose
388.35	0.7	CN	Magnetic network imaging
396.85	0.3	Ca II H	Chromospheric heating
430.50	0.8	CH	Magnetic elements
450.45	0.4		Blue continuum Temperature
555.05	0.4		Green continuum Temperature
668.40	0.4		Red continuum Temperature

Narrowband Filter Imager (NFI)

Field of view	328×164 arcsec (unvignetted 264×164 arcsec)
CCD	$4k \times 2k$ pixels (full FOV), shared with BFI
Exposure time	$0.1 - 1.6$ s (typical)
Spatial sampling	0.08 arcsec pixel^{-1} (full resolution)
Spectral resolution	0.009 nm (9 pm) at 630 nm

Spectral band (tunable filter)

Center (nm)	Width (nm)	Lines of interest	g_eff	Purpose
517.2	0.6	Mg I b 517.27	1.75	Chromospheric Dopplergrams and magnetograms
525.0	0.6	Fe I 524.71	2.00	Photospheric magnetograms
		Fe I 525.02	3.00	
		Fe I 525.06	1.50	
557.6	0.6	Fe I 557.61	0.00	Photospheric Dopplergrams
589.6	0.6	Na D 589.6		Very weak fields (scattering polarization) Chromospheric fields
630.2	0.6	Fe I 630.15	1.67	Photospheric magnetograms
		Fe I 630.2	2.5	
656.3	0.6	H I 656.28		Chromospheric structure

Standard observable examples for filter observations

Filtergram	A signal exposure photometric images	
	Frame size	$4k \times 2k$, $2k \times 2k$, $1k \times 2k$, or $0.5k \times 2k$
	Summing	1×1 ($1k \times 2k$ or smaller), 2×2, or 4×4 pixels
	Readout time	3.4 s (1×1 sum), 1.7 s (2×2), 0.9 s (4×4)
		Partial readout for faster cadence
	Reconfigure time	< 2.5 s (for changing filter wheels etc.)

Table 2 (*Continued*)

Dopplergram	Derived from narrowband filtergrams at several wavelengths	
	Frame size	4k × 2k, 2k × 2k, 1k × 2k, or 0.5k × 2k
	Summing	1 × 1 (1k × 2k or smaller), 2 × 2, or 4 × 4 pixels
	Duration	12.8 s (4 images, 2 × 2 sum, 0.8-s exposure)
Longitudinal magnetogram	Stokes V/I images converted onboard from narrowband filtergrams	
	Frame size	2k × 1k, 1k × 2k, or 2k × 2k
	Summing	1 × 1 (1k × 2k or smaller), 2 × 2, or 4 × 4 pixels
	Duration	8 images (4 wavelengths) are taken
		12.8 s for 1k × 2k and ≈ 21 s for 2k × 2k
Stokes *IQUV* (for vector magnetogram)	*IQUV* images made onboard from narrowband filtergrams at different polarization modulator positions		
	Shuttered exposures	Frame size	4k × 2k, 2k × 2k, 1k × 2k, or 0.5k × 2k
		Summing	1 × 1 (1k × 2k or smaller), 2 × 2, or 4 × 4 pixels
	Shutterless exposures	Frame size	Various
		Summing	1 × 1, 2 × 2, or 4 × 4 pixels
		Duration	1.6 – 12.8 s
			(1 – 8 waveplate rotations)

spectral lines span the photosphere to the lower chromosphere for diagnosis of dynamical behavior of magnetic and velocity fields in the lower atmosphere. The spectral bandwidth of the Lyot filter is ≈ 95 mÅ at 630 nm, and the wavelength center is tunable to several positions in a spectral line and its nearby continuum. There is no wavelength shift across the field of view because of the telecentric beam. It is noted that the edges of the full field of view are slightly vignetted owing to the limited size of the optical elements of the tunable filter residing in a telecentric beam. The unvignetted area is 264 arcsec in diameter. Exposure times are typically 0.1 – 0.4 s, but like the BFI, longer exposures are possible.

Filter observations mainly produce four types of observables: filtergrams, Dopplergrams, longitudinal magnetograms, and Stokes *IQUV* images. Filtergrams are snapshot images acquired from a single exposure for mapping the intensity of the solar features. "Broadband" filtergrams are the only observable made by BFI. "Narrowband" filtergrams are obtained by the NFI for all the spectral lines and nearby continuum located in the NFI spectral windows. The shutter open/close operations are always synchronized to the phase of the PMU. Various combinations of frame size and pixel summing mode may be chosen to reduce the data volume at the expense of FOV size and/or spatial resolution. The readout times for the full CCD are 3.4 s at 1 × 1 summing, 1.7 s at 2 × 2 summing, and 0.9 s at 4 × 4 summing. The readout of a smaller window of the CCD (several discrete sizes from 192 to 2048 rows) is possible in the central 2k × 2k pixel area for faster cadence as well as for reduced data volume. The time for reconfiguring mechanisms, including wavelength change by filter wheels, is less than ≈ 2.5 s. Onboard processing is performed in the FPP to make magnetograms, Dopplergrams, and Stokes parameters, and data compression is done in the MDP as described in the following.

Dopplergrams are images of the Doppler shift of a spectral line derived from narrowband filtergrams at several wavelengths. The central wavelength is derived from two or four images uniformly spaced through the line. Onboard memory processing is performed in real time in the FPP to calculate sums, differences, and ratios of images, which are sent to the ground separately. The data are converted on the ground to a velocity via a lookup table. The best photospheric line for Doppler measurements is Fe I 557.6 (Lande $g = 0$). The rms noise is typically 30 m s^{-1} for an observation with four images.

Longitudinal magnetograms give the location, the polarity, and a crude estimate of flux of the magnetic field components along the line of sight. Onboard processing in the FPP combines multiple narrowband filtergrams into two-image data (numerator and denominator) for reduced telemetry load. The primary lines are Fe I 630.25 and Fe I 525.02 (for the photosphere) and Mg I 517.27 (for the low chromosphere). Typical magnetograms take ≈ 20 s for eight images and have an rms noise of $\approx 10^{15}$ Mx per pixel.

Stokes *IQUV* images are made onboard from narrowband filtergrams at eight phases of the polarization modulator for each wavelength setting. Stokes demodulation is done to minimize noise caused by the time change of the Stokes *I*. Analysis of *IQUV* images at multiple wavelengths in a spectral line yields vector magnetic field information (*i.e.*, vector magnetograms). Shutterless modes with the frame transfer operation of the CCD are used for higher time resolution (1.6 – 4.8 s) and sensitivity, although the field of view is restricted by a focal plane mask. With 0.1-s exposure, 16 images are taken in a revolution of the PMU waveplate. These images are successively added or subtracted in the four slots of the smart memory to create the Stokes *IQUV* images. The modulation frequency is 2 per PMU rotation for *V* and 4 per PMU rotation for *Q* and *U*. Optionally, longer exposures may be used: with 0.4-s exposure (1/4 of the PMU rotation), we can measure only *V*; with 0.2-s exposure we can measure *Q, U,* or *V*; and with 0.1-s exposure, we can measure all *QUV*.

The processing in smart memory is identical to that for the SP (see Section 5.2). In shutterless mode, the FOV is 5.1 × 164 arcsec for 0.08-arcsec pixels, 12.8 × 164 arcsec for 0.16-arcsec pixels, and 25.6 × 164 arcsec for 0.32-arcsec pixels. Larger FOV values may be obtained by using successive exposures or longer exposure times (for partial Stokes sets). Stokes *IQUV* parameters also may be measured by using the mechanical shutter. The FOV is up to 82 × 164 arcsec for 0.08-arcsec pixels and 328 × 164 arcsec for 0.16- or 0.32-arcsec pixels. Up to 0.4-s exposures are possible for *V*, and up to 0.2-s exposures for *Q* and *U*. Note that additional noise sources resulting from the time between frames and crosstalk from Stokes *I* may appear.

5.2. Spectral Observations

The SP (Table 3) obtains line profiles of two magnetically sensitive Fe lines at 630.15 and 630.25 nm and the nearby continuum by using a 0.16 × 151 arcsec slit. Spectra are exposed and read out continuously 16 times per rotation of the PMU, and the raw spectra are added and subtracted onboard in real time to demodulate them, generating Stokes *IQUV* spectral images. Two spectra are simultaneously taken in orthogonal linear polarizations. When combined during the data analysis after downlink, this greatly reduces spurious polarization from any residual image jitter or solar evolution. The solar image may be stepped across the slit to map a finite area, up to the full 320-arcsec-wide FOV.

The SP is flexible in mapping observing regions, allowing one to perform suitable observations depending on science objectives. The SP only has a few modes of operation: Normal Map, Fast Map, Dynamics, and Deep Magnetogram. The Normal Map mode produces polarimetric accuracy of 0.1% with 0.15 × 0.16 arcsec pixels. It takes 83 min to scan a 160-arcsec-wide area: enough to cover a moderate-sized active region. By reducing the scanning size, the cadence becomes faster (50 s for mapping of a 1.6-arcsec-wide area), which would be useful for studying dynamics of small magnetic features. The Fast Map mode of observation can provide 30-min cadence for a 160-arcsec-wide scan with 030 × 0.32 arcsec pixel

Table 3 SOT/FPP Spectro-Polarimeter observations.

Spectro-Polarimeter (SP)		
Field of view along slit		163.84 arcsec (N – S direction)
Spatial scan range		327.62 arcsec (transverse to slit, E – W direction)
Spatial sampling (slit)		0.16 arcsec
Spectral line and coverage		Fe I 630.15 nm
		Fe I 630.2 nm
		Coverage: 630.08 to 630.32 nm
Spectral resolution/sampling		3pm/2.15pm
Measurement of polarization		Stokes *I, Q, U, V* simultaneously with dual beam (orthogonal linear components)
Polarization signal to noise		10^3 (normal map)

Standard observable (mapping mode) for the SP		
Normal mapping	Time per position	4.8 s (3 rotations of waveplate)
	Polarimetric accuracy	0.001
	FOV along slit	164 arcsec
	Sampling along slit	0.16 arcsec
	Data size	918k pixels in 4.8 s or 191k pixels s^{-1}
	Slit-scan sampling	0.16 arcsec
	Time for map area	50 s for 1.6 arcsec wide
		83 min for 160 arcsec wide
Fast mapping	Time per position	One rotation for the 1st slit position and another rotation for the 2nd slit position to form one slit data
	FOV along slit	164 arcsec
	Sampling along slit	0.32 arcsec
	Data size	459k pixels in 3.6 s or 127k pixels s^{-1}
	Slit-scan sampling	0.32 arcsec
	Time for map area	18 s for 1.6 arcsec wide
		30 min for 160 arcsec wide
Dynamics	Time per position	1.6 s (one rotation)
	FOV along slit	32 arcsec (to reduce data size)
	Sampling along slit	0.16 arcsec
	Data size	179k pixels in 1.6 s or 120k pixels s^{-1}
	Slit-scan sampling	0.16 arcsec
	Time for map area	18 s for 1.6 arcsec wide

size. In the Fast Map mode, the Stokes profiles at two slit positions with each integration time of 1.6 s are summed, and 2 pixels along the slit are also summed to give a polarization accuracy a factor of 1.15 better than 0.1%. The Dynamics mode of observation provides higher cadence (18 s for a 1.6-arcsec-wide area) with 0.16-arcsec pixels, although with lower polarimetric accuracy. In Deep Magnetogram mode, photons may be accumulated over many

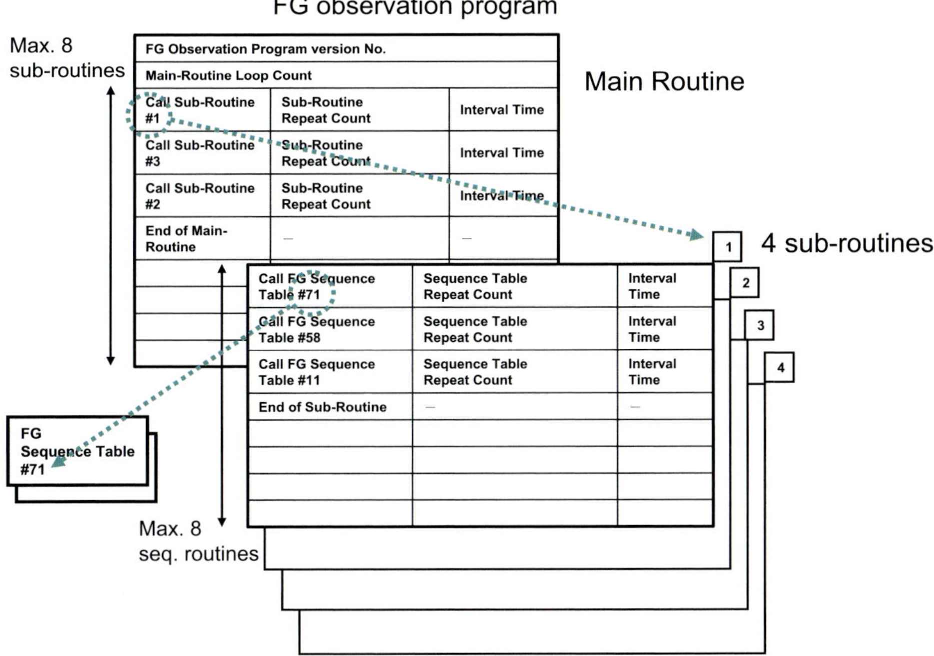

Figure 12 SOT Observation Program structure.

in total, of which ≈ 5.6 Gbits may be available for SOT data with an operational agreement of $\approx 70\%$ allocation to the SOT, although this allocation can be changed. Approximately 1.7 Gbits of SOT data can be downloaded through the nominal 4-Mbps high-telemetry channel in one ground station pass (assuming a 10-min duration). If 15 stations are scheduled per day, then the SOT can acquire 25.5 Gbits data per day, and the corresponding post-compression average data rate from the SOT is ≈ 300 kbps. An extreme example is to perform a high-rate burst observation, which provides high-cadence observations with wide field of view. The post-compression maximum data rate is ≈ 1.3 Mbps. When XRT and EIS observations are not solicited, which depends on the science purpose (the so-called SOT-dominant mode), the maximum rate can reach ≈ 1.8 Mbps. In the burst observation mode, it takes about one hour to completely fill the 5.6 Gbits of space in the spacecraft recorder. These data would require about three station passes for complete downlink.

8. Conclusions

The Solar Optical Telescope aboard *Hinode* is the largest aperture advanced solar telescope ever launched into space. The SOT consists of the Optical Telescope Assembly and the Focal Plane Package, and it obtains high-resolution photometric images from the photosphere to the chromosphere and makes highly accurate measurements of the vector magnetic fields with its filtergraph and spectropolarimeter. The stable cadence, unaffected by spacecraft night (*i.e.*, eclipses) or bad seeing, is particularly effective in obtaining high-quality movies, from which various discoveries are being made.

The in-orbit performance of the SOT is generally excellent and has met or exceeded all pre-launch expectations for the BFI, the SP, and the image-stabilization system. However, images from the NFI unfortunately contain artifacts that degrade or obscure the image over part of the field of view. These are caused by air bubbles in the index-matching fluid inside the tunable filter. They distort and move when the filter is tuned, and then usually drift toward the edges of the FOV over time. For this reason, NFI observing is usually done in one spectral line at one or a small number of wavelengths for extended periods of time. Rapid switching between lines is not allowed. Software changes made since launch have given us considerable control over the location of the bubbles; Targets can usually be placed in large blemish-free areas of the CCD. Tuning schemes have been developed that permit tuning to different positions in a line profile without disturbing the bubbles. This has enabled collection of most of the expected NFI observations. Flat-field correction of NFI images is still a challenge, but progress is being made on this; magnetograms and Dopplergrams are usually self-correcting since they are made from ratios of intensity differences. Details on the NFI performance will be published elsewhere.

Acknowledgements   The Solar Optical Telescope (SOT) aboard *Hinode* is the result of a fruitful international collaboration between Japan and the United States. The SOT design meetings were held 16 times, either in Tokyo or Palo Alto, until the start of the final-level spacecraft testing in August 2004. The extensive week-long interaction in the design meetings resulted in the successful design, fabrication, and joint tests of the mechanical, thermal, electrical, optical, and control/guidance components of the instrument. All of the participants in the program were impressed with the rapid development of what had initially appeared to be ambitious program to the sophisticated state-of-the-art instrument at its completion.

The principal investigator (PI) of the SOT is Saku Tsuneta. The U.S. PI for NASA had been Alan Title, and Ted Tarbell succeeded him in the fall of 2005. Needless to say, numerous scientists, engineers, and managers in Japan, the United States, and France contributed to the program. We sincerely thank those individuals and the organizations they represent.

The SOT consists of the Optical Telescope Assembly and the Focal Plane Package. The OTA and the image stabilization subsystem (CTM) of the SOT were built by the *Solar-B* project office and the Advanced Technology Center (ATC) of the National Astronomical Observatory of Japan (NAOJ) and their industry partners. The prime industry partner for OTA/CTM is the Communication Systems Center of the Mitsubishi Electric Cooperation, along with the participation of SAGEM REOSC (for primary, secondary, heat dump, and secondary field stop mirrors); Canon, Inc. (for CLU fabrication); Genesia (for CLU optical design and astigmatism corrector); Sankyo Optics Industry Co., Ltd., and Okamoto Optics Works, Inc. (for astigmatism corrector), Systems Engineering Consultants (SEC; for CTM flight software); Mitsubishi Space Software (for CTM digital electronics and OTA thermal design); Queensgate Instruments, Ltd. (for CTM actuators); and Mitsubishi Heavy Industries, Ltd. (for the MDP). Koichi Waseda of the NAOJ ATC designed and fabricated the flight IR filter for the CLU.

The FPP consists of a wide-band camera, narrow-band camera, spectropolarimeter, and correlation tracker. The FPP was built by the Lockheed Martin Advanced Technology Center, with participation from the High Altitude Observatory of the National Center for the Atmospheric Research (who are mainly responsible for the SP) and NAOJ. Significant subcontractors included E2V and Mullard Space Sciences Laboratory (for the CCD detectors), Barr Associates and Andover Corporation (for the filters), Vision Composites (for the structure), and Horber Magnetics (for the motors).

We would like to thank Hirohisa Hara, Tomonori Tamura, and Naoko Baba and her team at JAXA/ISTA (for contamination control), Ryohei Kano and Masahito Kubo (for the onboard Doppler correction algorithm), Ken Kobayashi (for CTM analog electronics), Kenji Minesugi (for structural work) and Akira Onishi (for thermal work) at JAXA/ISAS, and Keiichi Matsuzaki (for the MDP). (Personnel listed without affiliations are from NAOJ at the time of development.) We also thank Yasushi Sakamoto and Naoki Kohara for their contribution to the testing.

Izumi Mikami, Hideo Saito, Tadashi Matsushita, and Noboru Kawaguchi led the SOT program at the Communication Systems Center of the Mitsubishi Electric Corporation. Lead engineers of the Mitsubishi team include Toshitaka Nakaoji (OTA structure), Kazuhiro Nagae (OTA thermal), Yasuhiro Kashiwagi (CTM systems), Osamu Ito (CTM analog electronics), Yoshihiro Hasuyama (integration, inspection, and assembly of critical optics), Kazuhide Kodeki (CTM guidance and control), Masaki Tabata (CTM mechanism development), Norimasa Yoshida (guidance and control and microvibration), Tsuyoshi Ozaki (composite material), Nobuaki Kaido (OTA thermal), Shusaku Inoue (CTM digital electronics), and Jun Nakagawa (OTA deployment door).

We also thank Renaud Mercier Ythier, Luc Thepaut, Eric Ruch, and Daniel Mouricaud (SAGEM/REOSC), Hideo Yokota and Masaharu Suzuki (Canon), Masayuki Nagase (SEC), and Kim Streander (HAO) for their superb work.

Tim Gordon and Therese Errigo of Swales Aerospace contributed to various OTA contamination control issues, with support from the NASA Marshall Space Flight Center (Keith Albyn and Larry Hill) and the U.S. Naval Research Laboratory (Clarence Korendyke). Charles Powers of NASA Goddard Space Flight Center provided additional qualification for the DEB dampers of the OTA doors. Jim Bilbro and Scott Smith (NASA Marshall Space Flight Center) advised the OTA program in resolving some critical optical issues.

Michael Levay, Bruce Jurcevich, and Chris Hoffmann led the FPP development program in the United States. Department heads included Bill Rosenberg and Gary Kushner (systems engineering), Chris Hoffmann (assembly, integration and test), Dick Shine (tunable filter), David Elmore (SP), Chris Edwards (electrical, CT, and CCD cameras), Dnyanesh Mathur (software), Barbara Fischer (mechanical), Dave Akin (mechanisms), Ericka Sleight (thermal), and Tom Cruz (logistics). The *Solar-B* project office at NASA Marshall Space Flight Center extensively oversaw the SOT program in the United States. The NASA project office led by Larry Hill consisted of Jerry Owens, Robert Jayroe, Barbara Cobb, Vernon Keller, Danny Johnston, Charlotte Talley, and Spence Glasgow.

Sadanori Shimada and his team working in spacecraft systems (at Kamakura Works of the Mitsubishi Electric Corporation) supported the OTA development and SOT integration to the spacecraft.

S. T. would like to express deep appreciation to the former and the present director generals of the National Astronomical Observatory of Japan, Prof. Norio Kaifu and Prof. Shoken Miyama, for their strong support for the program. The authors thank Gary Kilper for comments on the paper.

References

Abbett, W.P.: 2006, *Astrophys. J.* **665**, 1469.

Bello Gonzalez, N., Okunev, O.V., Domínguez Cerdeña, I., Kneer, F., Puschmann, K.G.: 2005, *Astron. Astrophys.* **434**, 317.

Berger, T.E., Rouppe van der Voort, L.H.M., Löfdahl, M.G., Carlsson, M., Fossum, A., Hansteen, V.H., Marthinussen, E., Title, A., Scharmer, G.: 2004, *Astron. Astrophys.* **428**, 613.

Carlsson, M., Stein, R.F.: 2004, In: Sakurai, T., Sekii, T. (eds.) *The Solar-B Mission and the Forefront of Solar Physics*, ASP Conf. Ser. **325**, 243.

Cattaneo, F.: 1999, *Astrophys. J.* **515**, 39.

Cheung, M.C., Schüssler, M., Moreno Insertis, F.: 2007, *Astron. Astrophys.* **467**, 703.

Culhane, J.L., Harra, L.K., James, A.M., Al-Janabi, K., Bradley, L.J., Chaudry, R.A., *et al.*: 2007, *Solar Phys.* **243**, 19.

De Pontieu, B.: 2004, *Nature* **430**, 536.

Ferriz-Mas, A., Steiner, O.: 2008, *Solar Phys.* in press.

Fisher, G.H., Fan, Y., Longcope, D.W., Linton, M.G., Pevtsov, A.A.: 2000, *Solar Phys.* **192**, 119.

Gudiksen, B.V., Nordlund, A.: 2005, *Astrophys. J.* **618**, 1020.

Golub, L., DeLuca, E., Austin, G., Bookbinder, J., Caldwell, D., Cheimets, P., *et al.*: 2007, *Solar Phys.* **243**, 63.

Handy, B.N., Acton, L.W., Kankelborg, C.C., Wolfson, C.J., Akin, D.J., Bruner, M.E., *et al.*: 1999, *Solar Phys.* **187**, 229.

Hansteen, V.H., De Pontieu, B., Rouppe van der Voort, L., van Noort, M., Carlsson, M.: 2006, *Astrophys. J.* **647**, L73.

Harvey-Angle, K.L.: 1993, *Magnetic bipoles on the Sun.* Ph.D. Thesis, Utrecht University.

Ichimoto, K., Lites, B., Elmore, D., Suematsu, Y., Tsunete, S., Katsukawa, Y., *et al.*: 2008, *Solar Phys.*, in press.

Kano, R., Sakao, T., Hara, H., Tsuneta, S., Matsuzaki, K., Kumagai, K., *et al.*: 2007, *Solar Phys.*, in press.

Katsukawa, Y., Tsuneta, S.: 2005, *Astrophys. J.* **621**, 498.

Khomenko, E.V., Shelyag, S., Solanki, S.K., Vögler, A.: 2005, *Astron. Astrophys.* **442**, 1059.

Kosovichev, A.: 2004, In: Sakurai, T., Sekii, T. (eds.) *The Solar-B Mission and the Forefront of Solar Physics*, ASP Conf. Ser. **325**, 75.

Kosugi, T., Matsuzaki, K., Sakao, T., Shimizu, T., Sone, Y., Tachikawa, S., *et al.*: 2007, *Solar Phys.* **243**, 3.

Leighton, R.B.: 1964, *Astrophys. J.* **140**, 1547.

Leka, K.D., Metcalf, T.R.: 2003, *Solar Phys.* **212**, 361.

Lites, B.W.: 1996, *Solar Phys.* **163**, 223.

Lites, B.W., Leka, K.D., Skumanich, A., Martinez Pillet, V., Shimizu, T.: 1996, *Astrophys. J.* **460**, 1019.

Metcalf, T.R., Jiao, L., McClymont, A.N., Canfield, R.C., Uitenbroek, H.: 1995, *Astrophys. J.* **439**, 474.

Parker, E.N.: 1978, *Astrophys. J.* **221**, 368.

Parker, E.N.: 1988, *Astrophys. J.* **330**, 474.

Sekii, T.: 2004, In: Sakurai, T., Sekii, T. (eds.) *The Solar-B Mission and the Forefront of Solar Physics*, *ASP Conf. Ser.* **325**, 87.

Shimizu, T., Nagata, S., Tsuneta, S., Tarbell, T., Edwards, C., Shine, R., *et al.*: 2008, *Solar Phys.*, in press.

Scharmer, G.B., Gudiksen, B.V., Kiselman, D., Löfdahl, M.G., Rouppe van der Voort, L.H.M.: 2002, *Nature* **420**, 151.

Skartlien, R., Stein, R.F., Nordlund, A.: 2000, *Astrophys. J.* **541**, 468.

Solanki, S.K.: 2003, *Astron. Astrophys. Rev.* **11**, 153.

Solanki, S.K., Inhester, B., Schüssler, M.: 2006, *Pep. Prog. Phys.* **69**, 563.

Sterling, A.C.: 2000, *Solar Phys.* **196**, 79.

Suematsu, Y., Tsuneta, S., Ichimoto, K., Shimizu, T., Otsubo, M., Katsukawa, Y., *et al.*: 2008, *Solar Phys.*, in press.

Tarbell, T.D., *et al.*: 2008, *Solar Phys.*, to be submitted.

Title, A.M.: 2007, In: Shibata, K., Nagara, S., Sakurai, T. (eds.) *New Solar Physics with Solar-B Mission*, *ASP Conf. Ser.* **369**, 409.

Tsuneta, S.: 1996, In: Bentley, R.D., Mariska, J.T. (eds.) *Magnetic Reconnection in the Solar Atmosphere*, *ASP Conf. Ser.* **111**, 409.

Ulmschneider, P., Musielak, Z.: 2003, In: Pevtsov, A.A., Uitenbroek, H. (eds.) *Current Theoretical Models and Future High Resolution Solar Observations: Preparing ATST*, *ASP Conf. Ser.* **286**, 363.

Ulrich, R.K.: 1996, *Astrophys. J.* **465**, 436.

Walsh, R.W., Ireland, J.: 2003, *Astron. Astrophys. Rev.* **12**, 1.

Welsch, B.T., Abbett, W.P., DeRosa, M.L., Fisher, G.H., Georgoulis, M.K., Kusano, K., Longcope, D.W., Ravindra, B., Schuck, P.W.: 2007, *Astrophys. J.*, submitted.

Weiss, N.O.: 2006, *Space Sci. Rev.* **124**, 13.

Withbroe, G.L., Noyes, R.W.: 1977, *Ann. Rev. Astron. Astrophys.* **15**, 363.

Yoshida, T., Tsuneta, S.: 1996, *Astrophys. J.* **459**, 342.

The Solar Optical Telescope of *Solar-B* (*Hinode*): The Optical Telescope Assembly

Y. Suematsu · S. Tsuneta · K. Ichimoto · T. Shimizu · M. Otsubo · Y. Katsukawa · M. Nakagiri · M. Noguchi · T. Tamura · Y. Kato · H. Hara · M. Kubo · I. Mikami · H. Saito · T. Matsushita · N. Kawaguchi · T. Nakaoji · K. Nagae · S. Shimada · N. Takeyama · T. Yamamuro

Originally published in the journal Solar Physics, Volume 249, No 2.
DOI: 10.1007/s11207-008-9129-4 © Springer Science+Business Media B.V. 2008

Abstract The Solar Optical Telescope (SOT) aboard the *Solar-B* satellite (*Hinode*) is designed to perform high-precision photometric and polarimetric observations of the Sun in visible light spectra (388–668 nm) with a spatial resolution of 0.2–0.3 arcsec. The SOT consists of two optically separable components: the Optical Telescope Assembly (OTA), consisting of a 50-cm aperture Gregorian with a collimating lens unit and an active tip-tilt mirror, and an accompanying Focal Plane Package (FPP), housing two filtergraphs and a spectro-polarimeter. The optomechanical and optothermal performance of the OTA is crucial to attain unprecedented high-quality solar observations. We describe in detail the instrument design and expected stable diffraction-limited on-orbit performance of the OTA, the largest state-of-the-art solar telescope yet flown in space.

Keywords Sun: instrumentation · Sun: space telescope · Sun: visible light

Y. Suematsu (✉) · S. Tsuneta · K. Ichimoto · M. Otsubo · Y. Katsukawa · M. Nakagiri · M. Noguchi · T. Tamura · Y. Kato · H. Hara
National Astronomical Observatory of Japan, 2-21-1 Osawa, Mitaka, Tokyo 181-8588, Japan
e-mail: suematsu@solar.mtk.nao.ac.jp

T. Shimizu · M. Kubo
Institute of Space and Aeronautical Science, JAXA, 3-1-1 Yoshinodai, Sagamihara,
Kanagawa 229-8510, Japan

Present address:
M. Kubo
High Altitude Observatory, National Center for Atmospheric Research, P.O. Box 3000, Boulder,
CO 80307, USA

I. Mikami · H. Saito · T. Matsushita · N. Kawaguchi · T. Nakaoji · K. Nagae
Communication Systems Center, MELCO, 8-1-1 Tsukaguchi-Honmachi, Amagasaki, Hyogo 661-8661,
Japan

S. Shimada
Kamakura Works, MELCO, 325 Kami-Machiya, Kamakura, Kanagawa 247-8520, Japan

N. Takeyama · T. Yamamuro
Genesia Corporation, 3-38-4-601 Shimo-Renjaku, Mitaka, Tokyo 181-0013, Japan

1. Introduction

The aim of the Solar Optical Telescope (SOT) aboard the *Solar-B* satellite (postlaunch named *Hinode*) is to provide high-precision photometric and polarimetric data to investigate magnetic origins and mechanisms of active phenomena on the Sun. Additionally, the SOT is designed to explore the physical coupling between the photosphere and the upper layers to understand the mechanism of dynamics and heating with the help of the coordinated observations from the X-Ray Telescope (Golub *et al.*, 2007; Kano *et al.*, 2008) and EUV Imaging Spectrometer (Culhane *et al.*, 2007) flown on *Hinode* (Kosugi *et al.*, 2007).

Owing to the Sun-synchronous polar orbit of *Hinode*, the SOT is expected to be able to continuously observe solar atmospheric structures, especially solar magnetic structures, with a diffraction-limited resolution and a polarization accuracy better than 10^{-3}. It is difficult for ground-based solar instruments to stably achieve these levels of performance, because the magnetic fields are concentrated in subarcsec structures that are much smaller than the atmospheric seeing-limited resolution. Magnetic components transverse to the line of sight, which give a measure of excess magnetic energy, are particularly difficult to observe and cannot be measured with any degree of accuracy if the magnetized structure is not spatially resolved. It should be stressed that observations from space have advantages not only in their capability of providing continuous coverage and high spatial resolution but also in offering wide field of view coverage and more stable intensity levels than those of ground-based observations.

The SOT was designed to meet the following basic specifications: It should observe the field of view fully, covering a moderate-sized active region of $\approx 3 \times 5$ arcmin wide, with a spatial resolution corresponding to small-scale magnetic elements of $0.2-0.3$ arcsec and with negligibly small and/or well-calibrated instrumental polarization. The SOT comprises very sophisticated instruments and consists of two optically separable components: the Optical Telescope Assembly (OTA) and the Focal Plane Package (FPP). This paper will focus on the OTA instrument design and its diffraction-limited performance expected on orbit. A series of accompanying papers will describe other key components in detail: Tsuneta *et al.* (2008) for the overview of the SOT, Tarbell *et al.* (2008) for the FPP, Shimizu *et al.* (2004, 2008) for the image stabilization system of the SOT, and Ichimoto *et al.* (2004, 2008) for the instrumental polarization calibration of SOT.

2. OTA Instrumentation

The SOT was designed to achieve the diffraction limit as a whole system. Following a conventional definition of the diffraction limit (Maréchal criterion; see, *e.g.*, Schroeder (2000) and Wilson (1996)), we defined the goal of the SOT having a Strehl ratio larger than 0.8 at 500 nm at the center of the field view, assuming evenly budgeted Strehl ratios of 0.9 for both the OTA and the FPP. The Strehl ratio is the peak intensity of a point source formed by a telescope normalized with the peak formed by a perfect telescope of no wavefront aberration. The Strehl ratio (SR) can be expressed with a root-mean-square (rms) wavefront error (WFE) by the relation

$$\text{SR} = \exp\left[-(2\pi\,\text{WFE}/\lambda)^2\right],$$

and then the budget of the OTA is 25.8 nm rms WFE, whereas the SOT has 36.5 nm rms. To achieve this goal, the budget was subdivided for image-forming components and controlled during their fabrications and tests. The budget was also allocated for wavefront errors of

OTA optomechanical and optothermal origins. In addition to optical performance, the effect of pointing jitters was also included in the budget. Image blurring from jitter causes degradation of peak intensity, which can be expressed with WFE by pretending that the peak degradation corresponds to the Strehl ratio.

There are two major sources of pointing jitter: one of lower frequency, originating from spacecraft body-pointing excursions and another of higher frequency and of OTA pointing-axis fluctuations (microvibrations of the $M1 - M2$ alignment) induced by moving elements in the attitude-control system and in observation instruments such as shutters, filter wheels, moving mirrors, and so on. To suppress the low-frequency pointing jitter, an active image-stabilization system consisting of a correlation tracker (in the FPP), tip-tilt mirror, and satellite pointing jitter controller was designed and developed for the SOT (see Shimizu *et al.*, 2004, 2008). A bench test of this system demonstrated a superb performance of 0.002 arcsec rms stabilization with a cross-over frequency of 14 Hz.

2.1. Optical Design and Components

Like most previous designs for large-sized space solar telescopes (Dunn, 1981; Title, 1989), the basic optical design of the OTA was determined to be Gregorian: an axisymmetric primary and secondary mirror system. One advantage of the Gregorian system is that field stops can be set at the primary focus and secondary focus (a Gregorian focus) to reject unwanted solar light to space. With the field stop at the primary, the heat load to the secondary mirror and downstream optics can be considerably reduced, as explained in Section 2.4. The uniqueness of the OTA, however, comes from a collimating lens unit placed near the center of the primary mirror, with which a compact telescope was achieved, which resulted in very smooth integration and testing.

The OTA was designed to fulfill the following scientific and engineering requirements: (1) to resolve at least 0.2-arcsec solar features over a field view of 320×160 arcsec, (2) to have a negligible chromatic aberration with a wide coverage of observation wavelengths from 388 to 668 nm without focus adjustment and to give a well-defined optical interface with the FPP, (3) to give negligible instrumental polarization before a polarization modulator for precise polarization measurements, and (4) to accommodate thermal design to reject unwanted solar light from the telescope components as early as possible. Accommodating the requirements of high spatial resolution and simultaneous wide-field coverage leads to a 50-cm telescope aperture, which can give a theoretical resolution of better than 0.2 arcsec at wavelengths shorter than 500 nm onto a 4000×2000 pixel detector at the FPP. This half-meter aperture size also met the limited resources available, such as those of payload launch capacity (of the JAXA M-V rocket) and test facilities.

The collimator lens, which can fulfill requirements (2) and (3), was designed to be placed near the center of the primary mirror and to reduce beam size, making an exit pupil of 30 mm diameter to accommodate the clear apertures of the following polarization modulator and an active tip-tilt mirror for image stabilization. An afocal beam from the collimator lens is also of benefit to relax the positional tolerance for the FPP with respect to the OTA.

After considerable optomechanical tradeoff studies involving the allowable size of the launcher's nosecone (whose maximum length ≈ 2 m), a distance of 1500 mm between the primary and the secondary mirrors was chosen. Longer telescopes have the advantage of allowing the relaxation of the positional tolerance of the secondary mirror; however, shorter telescopes can give better mechanical stability and also require a smaller secondary mirror (central obscuration) and hence offer better image contrast for observing solar granulation.

The idea was that the OTA should be a Gregorian with a collimating lens unit (CLU) near the center of the primary mirror, followed by a polarization modulator unit (PMU), and an active tip-tilt mirror (CTM-TM). In addition, the OTA has two field stops in between the primary and secondary mirrors: a heat dump mirror (HDM) at the focus of the primary mirror and the a secondary field stop (2FS) at the Gregorian focus. In addition, the following more practical requirements finally determined the basic optical parameters of the OTA:

1. The Gregorian should be aplanatic (both spherical and coma aberration free) to give better image quality over the specified wide field of view (FOV).
2. The entrance pupil was positioned 200 mm in front of the secondary mirror vertex.
3. The principal point of the CLU was positioned 50 mm beneath the vertex of the primary mirror and the 30-mm-diameter exit pupil formed 300 mm below the CLU, around which the moving optical elements, the PMU and CTM-TM, were located.
4. The HDM outer diameter is about twice the diameter of the solar image at the primary focus so that it allows an offset pointing of the telescope for solar limb observations up to 200 arcsec off the limb.
5. The HDM has a through hole passing the beams with a field of view of 500-arcsec diameter (400 arcsec plus 0.3 mm margin) and it should not vignette any beams reflected back from the secondary mirror with a clear margin of at least 0.5 mm.

The derived optical parameters are given Table 1 and the optical layout of the OTA is shown in Figure 1. In the following, details of each optical component are described.

M1 and M2   The primary (M1) and secondary (M2) mirrors for the aplanatic Gregorian were made of lightweight ULE (Figure 2). The surface figures of the mirrors were ellipsoids with conic constants shown in Table 1. Since null correctors (CGH) for these aspheric mirrors can contribute to inaccurate surface figures, we included an M1 – M2 combination test in the mirror-polishing process. As a result, touch-up polishing was needed for M1 to achieve the budget of 19.8 nm rms for the M1 – M2 Gregorian system, after polishing individual mirrors at the level of 12 nm rms wavefront using null correctors. Although the resulting vertex radii and conic constants were a bit changed from the original design (Table 1), the wavefronts were superior not only at field center but also at the extreme four field corners, achieving less than 18 nm rms. A space-qualified protected silver coating was deposited onto both mirrors for high throughput in the observation wavelengths and low solar absorption of solar light (Figure 7).

CLU   The CLU was one of the most challenging optical components to design. It had to be fabricated to fulfill stringent requirements, including radiation-hardened optics and structure, high throughput in the UV (down to 388 nm), a temperature-insensitive focus position, the ability to compensate for the large field curvature of the Gregorian, and negligibly small chromatic aberration in the observation wavelength range. Since we could use a silica radiation-insensitive optical glass as a radiation shield, we examined possible combinations of a silica lens element with two other glass materials for a temperature-insensitive apochromat. Millions of combinations of optical glasses were studied by using existing worldwide catalog data (CTE, n, dn/dT). Eventually, we found that one apochromat was not enough, but the combination of two apochromats having opposite signs of low-temperature sensitivity should work. The adopted glasses and layout of the optical design are shown in Figure 3.

Each lens was mounted with an accuracy of a few microns into a barrel of pure titanium. Testing of an as-built CLU confirmed a negligibly small focus position temperature sensitivity of 1.7 μm K^{-1}, with a corresponding wavefront change of 1.3 nm rms. The chromatic aberration is also small, as given in Table 1, confirming that the designed and measured values are consistent. It should be stressed that the wavefront of the CLU is superior, 9.3 nm rms in the field center, owing to elaborate iterations between the design and fabrication process.

Table 1 Parameters of OTA optical components.

Component	Design parameters	Remarks
Entrance pupil		Serves as sunshade
Position	200 mm ahead of M2 vertex	
Outer diameter (mm)	500	
Inner diameter (mm)	172	Three spiders 40 mm wide
Central obscuration	0.344	$= 172/500$
Primary mirror (M1)		ULE
Outer diameter (mm)	560	
Clear aperture (mm)	509	
Vertex radius (mm)	2339.4 ± 2.5	2342.98, measured
Conic constant	-0.9706 ± 0.001	(-0.9726, estimated)
Secondary mirror (M2)		ULE
Outer diameter (mm)	159	
Clear aperture (mm)	147	
Vertex radius (mm)	524.94 ± 2	525.04, measured
Conic constant	-0.3996 ± 0.001	(-0.39747, estimated)
M1 – M2 distance (mm)	1500 ± 3	CFRP truss
Gregorian focal length (mm)	4527 ± 25	
Collimator lens (CLU)		
Position	To give collimated exit beam in air	
Focal length (mm)	271.64 ± 0.3	
Chromatic aberration	$< 35\ \mu m$	For $\lambda = 388 - 668$ nm
Exit pupil		
Outer diameter (mm)	30 ± 0.3	
Distance from M1 (mm)	409 ± 5	
Field stops		
Heat dump mirror (HDM)		Aluminum alloy substrate
Position	M1 focus	
Mirror	$45°$ tilt flat	
Outer diameter (mm)	32.83	
Through hole FOV	$\phi = 505$ arcsec	
Secondary field stop (2FS)		Aluminum alloy substrate
Position	Gregorian focus	
Mirror	Cone with conical angle of $173°$	
Outer diameter (mm)	65	
Through hole FOV	361.3×197.4 arcsec	Rectangular hole

In the six-lens design, the two silica lenses were placed sunward so that they work as a shield against radiation mostly coming from the entrance. The first lens is meniscus shaped and the radius of curvature of its Sun-side surface coincides with the distance to the 2FS.

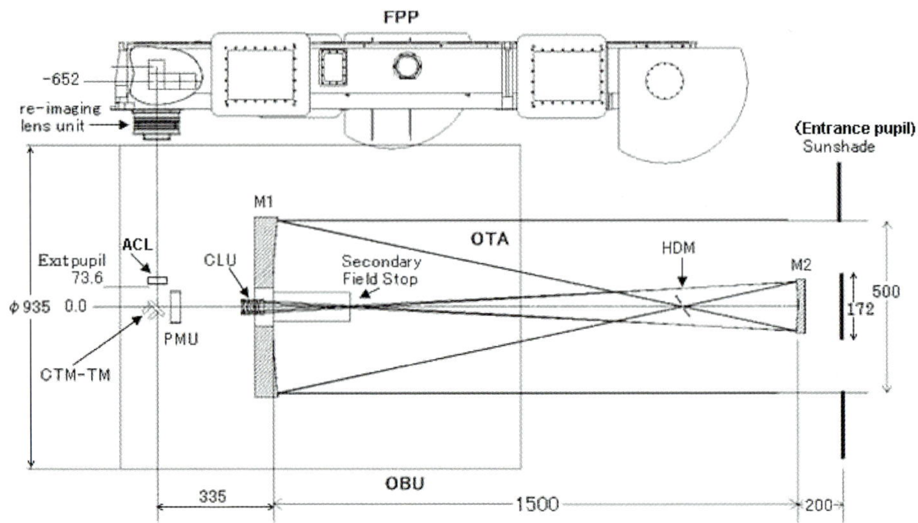

Figure 1 Optical configuration of OTA. Units are in millimeters. The OTA consists of an aplanatic Gregorian, a primary mirror (M1) and a secondary mirror (M2) of effective aperture of 500 mm, a collimating lens unit (CLU) near the center of the primary mirror, a polarization modulator unit (PMU), an active tip-tilt mirror (CTM-TM), and an astigmatism corrector lens (ACL). In addition, the OTA has two field stops between the primary and secondary mirrors; one is a heat dump mirror (HDM) at the focus of the primary mirror and the other is a secondary field stop (2FS) at the Gregorian focus.

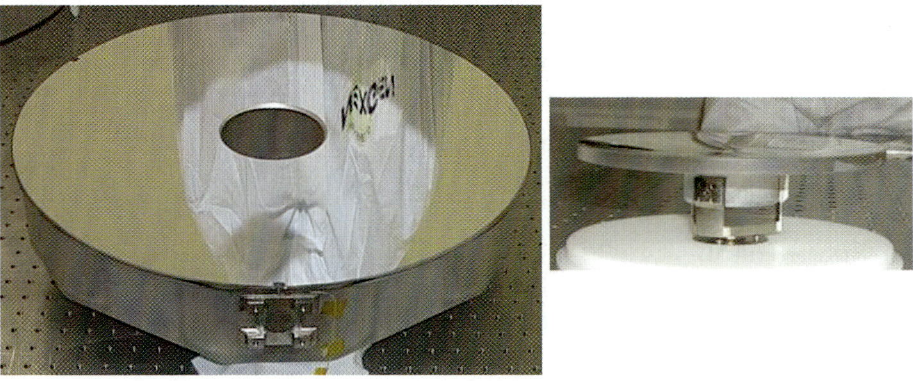

Figure 2 The flight-model primary mirror (left) and secondary mirror (right).

Therefore, unwanted light incident upon the CLU is reflected sunward through 2FS, M2, M1, and the entrance pupil. The dedicated IR/UV rejection multilayer coating, which was developed and extensively tested for space use by NAOJ, was deposited on the first surface. Other surfaces have a space-qualified antireflective (AR) coating and the measured total throughput is shown in Figure 7.

However, it turned out that the CLU has a non-negligible temperature-dependent linear retardation (Ichimoto *et al.*, 2008). Experiments show that the retardation of the CLU can be regarded as uniform over the field of view and constant against temperature if its temperature

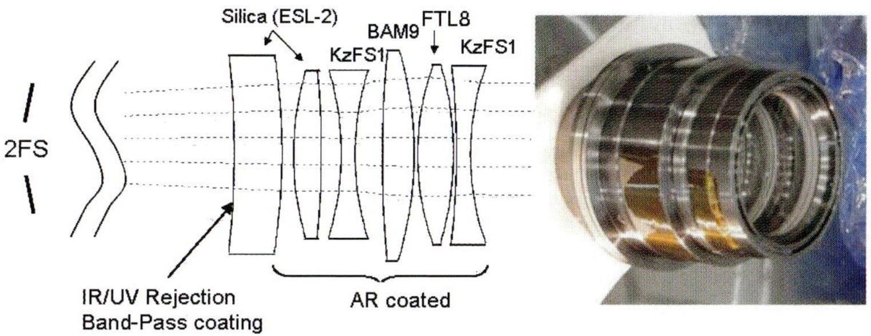

Figure 3 Layout of CLU lenses (left) and an aft view of flight-model CLU (right).

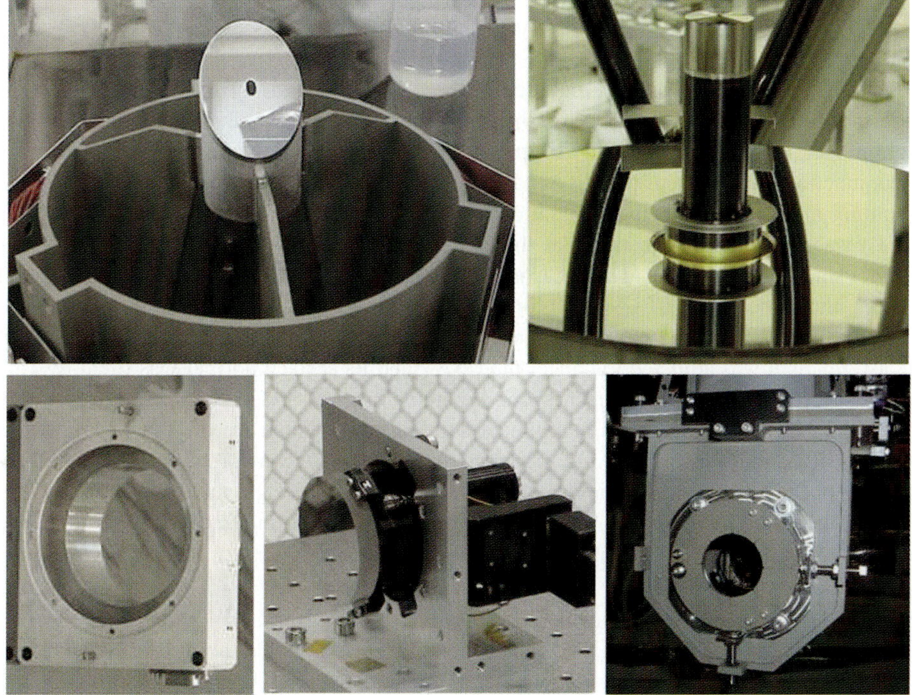

Figure 4 Photographs of the HDM welded at inner spiders by electron beam welding (upper left), 2FS installed at mirror cell (upper right), PMU in titanium housing (lower left), CTM-TM set for wavefront measurements (lower middle), and ACL assembled at the exit of the CTM-TM housing (lower right).

is higher than 25°C, which can be achieved through the control of dedicated operational heaters.

PMU   The PMU (Figure 4) is a rotating waveplate to modulate the intensity on the CCD according to the polarization of incident light. The PMU is located near the pupil image formed by the CLU for minimal instrumental polarization. The waveplate is a bicrystalline retarder consisting of two plates, one made of quartz and the other of sapphire, acting as a matched pair (Guimond and Elmore, 2004). It is noted that the waveplate is made of radiation-resistant and nonoutgassing materials. The retardation was optimized for linear and circular polarization at 630 and 517 nm. The waveplate is continuously rotated with a period of 1.6 s (nominal) by a DC hollow-core motor. Each wedge of the waveplate was designed to be very small and yield negligibly small image wobbling during its rotation (less than 0.2 arcsec on the Sun). This small image wobbling from the modulator is slow (0.625 Hz) enough to be canceled on the CCD by the correlation tracker system.

CTM-TM   This mirror is located near the exit pupil near the CLU to fold the beam to the FPP and to actively stabilize the image jitter at frequencies less than 14 Hz. It is made of ULE and is 60 mm round and 10 mm thick. Since the mirror is used to reflect the beam at an angle of 45°, its optical performance was evaluated in the same configuration. The resultant 4.5 nm rms wavefront error of this mirror is negligibly small when measured in the 32-mm clear aperture. The mirror has an enhanced silver coating optimized at 45° incidence so that its reflectivity is maximum or diattenuation is minimum (for the reflectivity, see Figure 7).

ACL   This component was not included in the original design but later turned out to be necessary for the OTA to eliminate non-negligible primary astigmatism found in an as-built flight model. It was likely that the astigmatism was caused by stress from a bolt connection in conjunction with uneven spacing of adhesive between M1 pads and interface plates of the mounting mechanisms. To recover the optical performance, an astigmatism corrector lens (ACL) was designed and installed at the exit pupil of the OTA (at the exit of the CTM-TM housing box; Figure 4). The ACL is a nearly plane parallel plate made of fused silica (ESL-2, the same as the CLU silica), is 60 mm in diameter and 10 mm thick, and has the simple astigmatism of 0.074λ at 632.8 nm in a central 30-mm-diameter area. The angle of the astigmatism was adjusted to cancel the OTA's and its surface normal was tilted by 1.5° with respect to the optical axis to prevent any ghost images caused by reflections returned from the FPP. The ACL has AR coatings at both surfaces, which are the same as those for the CLU. The addition of the ACL greatly reduced the OTA wavefront error and made it capable for the first time of achieving its diffraction-limited design goal by a large margin. It should be noted that the ACL has been present on the OTA during all flight model tests described here.

HDM and 2FS   The HDM is located at the primary focus and is a 45° flat mirror with a central hole of $\phi = 505$ arcsec. The HDM was designed to reflect about 90% of incident solar energy out to space through a window at the side of the OTA (a heat dump window). The outer diameter of the HDM, which is about twice the diameter of the solar image at the primary focus, determines the maximum offset pointing angle from the Sun center allowable for the spacecraft. The secondary 2FS is placed at the Gregorian focus for the purpose of further reduction of energy sent to the following optics. The 2FS defines the field of view of the OTA as 360×200 arcsec, which somewhat oversizes the area of the CCD in the FPP. Both mirrors were made of aluminum alloy and welded by electron beam welding to their support structures of the same material so that the heat absorbed by the mirror can spread over the support structure by thermal conduction and be emitted by large-area radiation. Both mirrors have an enhanced silver coating whose overall reflectivity is higher than the protected silver coating. The coating was extensively tested and verified for space and solar use; it completely cleared all the necessary tests for radiation, UV irradiation, thermal cycling between $-40°C$ and $80°C$, high humidity, heat spots, intense solar-light illumination, and adhesive tape.

2.2. Mechanical Design

The framework structure of the OTA should be lightweight but sufficiently robust to support and maintain the optical elements with a required positional accuracy against violent launch environmental conditions and severe on-orbit thermal conditions without any dedicated alignment mechanisms. The SOT has a single focusing (reimaging) mechanism at the entrance of the FPP (Figure 1).

The Gregorian OTA requirements demand very small static misalignment tolerances for the primary and secondary mirrors, on the order of a few tens microns for decenter and despace or several arcsec for tilt, and a micron-order despace short-term stability on orbit during observations (see Table 2). To meet this requirement, the telescope framework was made of a truss of newly developed ultra-low-expansion carbon-fiber-reinforced plastic (CFRP) pipes in a graphite cyanate matrix (Ozaki *et al.*, 1996). The CTE was proven to be smaller than ± 0.1 ppm K^{-1}, and the dimensional change from moisture absorption was measured to be about 30 ppm, which is much smaller than conventional epoxy matrix composite pipes. Three CFRP honeycomb sandwich panels (rings) were adhesively bonded with upper and lower truss pipes without any metal junctions to save weight and also to avoid differential CTE, which might cause unexpected telescope thermal distortion.

The overall layout of the OTA structural assembly is shown in Figure 5. The center panel ring (called the center section) provides the mechanical interface to the spacecraft; the OTA is mounted on the CFRP-made cylindrical optical bench unit (OBU) to the spacecraft with three quasi-kinematic titanium alloy mounting legs with stress-relief spring structures. The center section is equipped with alignment cubes at the top surface, which represent the mechanical and optical axes of the OTA, and are used for co-alignment among other telescopes and spacecraft attitude-control-system sun sensors.

Mounting of the primary mirror is one of the most critical parts of the OTA. The primary mirror, made of lightweight (70% removed and thus a mass of 14 kg) ULE, is supported by three stress-free mounting mechanisms seated on the CFRP bottom panel (called the mirror cell), interfaced with three superinvar pads bonded on the side of the mirror (Figure 2). The pad interface of the mounting mechanism is torque-free about three axes and also free in the radial direction, thus providing a kinematic mount for the primary mirror. The pad interface thus avoids stresses to the mirror resulting from dimensional errors in machining or temperature change. The only significant surface error of the primary mirror is caused by the difference of CTE between the superinvar pads and the ULE, which constrains the best-performance temperature range of the primary mirror to be between $-15°C$ and $55°C$.

Table 2 Misalignment sensitivity of OTA optical components to wavefront degradation.

Component	Misalignment	Wavefront error (nm rms)	Main aberration
Primary mirror	Despace	9.5/50 μm	defocus and spherical
	Decenter	2.8/10 μm	coma
	Tilt	16.1/10 arcsec	coma
Secondary mirror	Despace	10.1/50 μm	defocus and spherical
	Decenter	2.8/10 μm	coma
	Tilt	4.3/10 arcsec	coma
CLU	Despace	8.8/500 μm	defocus and spherical
	Decenter	0.5/500 μm	coma
	Tilt	1.1/30 arcmin	coma

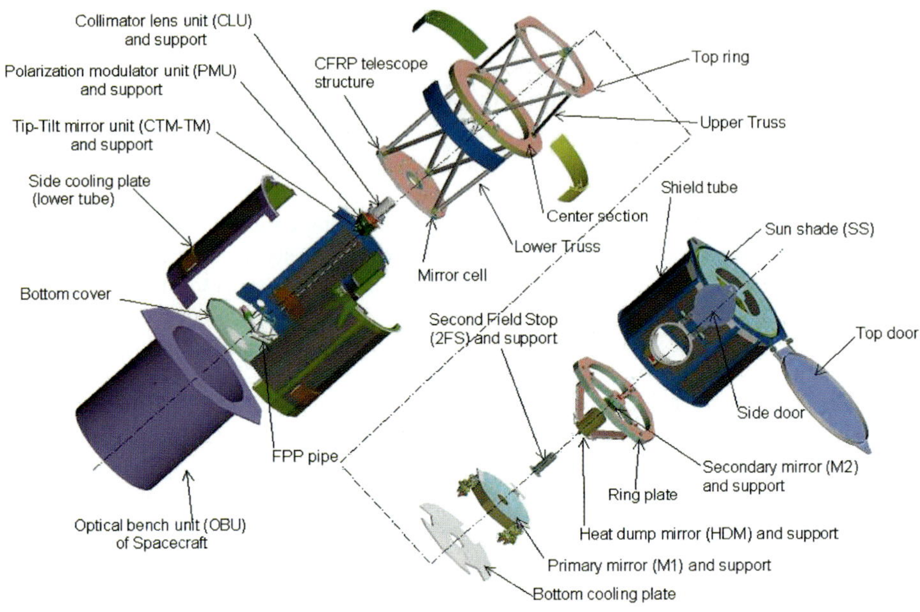

Figure 5 Layout of the OTA structural assembly. The OTA is mounted on an optical bench unit (OBU) at its center section by three titanium legs. The total weight of the OTA is 103 kg.

Titanium cylinders housing the CLU and PMU and supporting the CTM-TM aft end are also tightly mounted on the mirror cell, whose positional tolerances are relatively loose, after the secondary mirror is optically aligned with the primary.

The secondary mirror, which is also made of ULE shaped like a large Japanese flat wine cup, is supported by a superinvar tripod-shaped pad of stress relief spring legs, which are glued at the side of the M2 backside cylindrical hump (Figure 2). The pad located on the opposite side of M2 is bolted to the central part of another CFRP panel ring with spiders (called the ring plate). The surface error of the secondary mirror can again be caused by the CTE difference between the superinvar pad and the ULE, and it constrains the best-performance temperature range to between 0°C and 40°C.

The HDM unit is also supported by the ring plate via three mounting spiders made from honeycomb panel (aluminum core and CFRP skin). The mounting points at the ring plate have a titanium-made mechanism for decenter and tilt adjustment of the HDM unit with respect to M2. The HDM has a through hole through which the HDM can be properly aligned with a fiducial at the M2 vertex. The ring plate is connected to the CFRP top panel ring via three positional and tilt adjustment mechanisms made of superinvar rods, with which optical alignment of the secondary mirror, decenter, despace (focus), and tilt adjustment with respect to the primary mirror can be performed.

The OTA is covered by a shield tube for the upper half and a lower tube for the lower half for the purpose of protecting critical components from molecular and particle contamination, as well as reducing stray light, and ensuring thermal control of the entire OTA. The shield tube is made of aluminum honeycomb sandwich plates and also provides a structural support of top and side doors, along with a sunshade defining the entrance pupil of OTA.

The lower tube is made of thin aluminum plate; operational and survival heaters for optics are attached to its bottom cover.

Figure 6 Photographs showing the OTA truss structure and its optical components installed (left) and the final form of the OTA before the top and side door assembly (right).

The exiting ray path reflected by the CTM-TM toward the FPP is covered by an aluminum cylinder to avoid stray light. The lower tube and the OBU have a side hole through which we can insert a folding mirror or beam splitter into the exit beam path; this injection port is prepared to enable the optical tests of both the OTA and the FPP after their integration into the spacecraft during the final testing phase before launch.

The OTA has two doors: a top door, which covers the entrance pupil, and a side door, which covers the heat dump window. These doors are designed to deploy once on orbit. The role of the doors is to protect the OTA optics from dust particles during prelaunch tests and also liftoff. Additionally, they serve as cold traps for contaminants outgassing from structures on orbit before the deployments. Each door is held with a SUS strand and two hinges at the shield tube, and they can be deployed by a wire cutter and the force of two spiral springs at the hinges. Door deployment speed is suppressed by a space-qualified viscous damper at the hinge so that the deployment does not greatly disturb the satellite attitude. The door latches onto the hinges when it opens by 180°. The temperature of the wire cutters and the dampers are maintained with survival heaters, which can be enabled soon after launch.

Figure 6 shows the OTA when all the optical components were installed on the frame structure (left panel) in its final form covered with the shield tube and the lower tube (right panel) just before the top and side doors were assembled.

2.3. Contamination Control

Because contamination of optical surfaces and resulting thermal deformation and degradation of photon through put are among the issues most critical to the success of the SOT, we paid much attention to all the contaminant-doubtful materials and adhesives and spent much time in selecting low-outgassing materials. Before OTA integration, we performed grease removal cleanings and thorough thermal vacuum bakeout of not only most structure flight components but also of nonflight tools except for optical components. Four witness mirrors – two of them facing M1 and the others facing M2 – were inserted at a port in the center section to monitor contamination during the entire period through optical tests, thermal vacuum tests, optothermal tests, integration, and up to launch. We confirmed that the reflectivity of the witness mirrors at the wavelength of hydrogen Lyman alpha had never changed. All the activities of OTA integration and optical tests were carried out in a dedicated clean booth of class 100 or less installed in a cleanroom. Dust particle counts were continuously monitored to maintain required levels throughout all integration and testing.

2.4. Thermal Design

About 210 W of solar light is inevitably impinged onto the primary mirror at the bottom of the OTA during solar observations from its 500-mm-diameter entrance aperture. The lower half of the OTA is inserted into the cylindrical OBU, which provides a stable and isotropic thermal environment for the primary mirror. As a result, there is no short path to dump the heat absorbed by the primary mirror to space.

A thermal design to dump such a large heat load to space and maintain critical optical and structure components to within allowable temperature ranges with small temperature fluctuation is critically important to realize a high-performance solar telescope. The operational (best-performance) primary and secondary mirror temperature ranges cited earlier are required to maintain on-orbit OTA performance. From this viewpoint, the coating design of optical components is critical; these should limit solar light absorptance to a minimum, giving high throughput in the observation wavelengths and rejecting light outside their wavelengths (IR and UV).

The protected silver coating of M1 and M2 contributed a small solar absorptance of 6.5%, and the enhanced silver coating of the HDM and the 2FS contributed a lesser solar absorptance of 6.1%; actual absorptances for the HDM and 2FS are 3.9% after M1 and 3.6% after M1 plus M2 reflection, respectively. The first surface of the CLU has a multilayer coating for IR/UV rejection. The shape of the first surface of the CLU is concave with its center of curvature coinciding with the center of the secondary field stop. Thus, the CLU acts as an IR-blocking filter with the reflected light through the secondary field stop. It is noted that the major fraction of rejected light by the secondary field stop and the CLU can escape to space through the entrance aperture of the OTA. The CTM-TM mirror has an optimized enhanced silver coating for a 45° incidence and hence has small solar absorptance.

Based on the predicted orbit of *Hinode*, extreme cases were defined and studied for OTA thermal design: the "cold case" (solar limb observation in the coldest orbit with measured absorptance at the beginning of life) and the "hot case" (solar disk center observation in the hottest orbit with assumed absorptance increase by 5% toward the end of life). The basic concept of the OTA thermal design can be summarized as follows (see Figure 8):

1. Most incident energy (165 – 185 W) coming inside the OTA is reflected back by the primary mirror and dumped out to space by the HDM at the primary focus through the heat dump window opened at the side of the OTA.

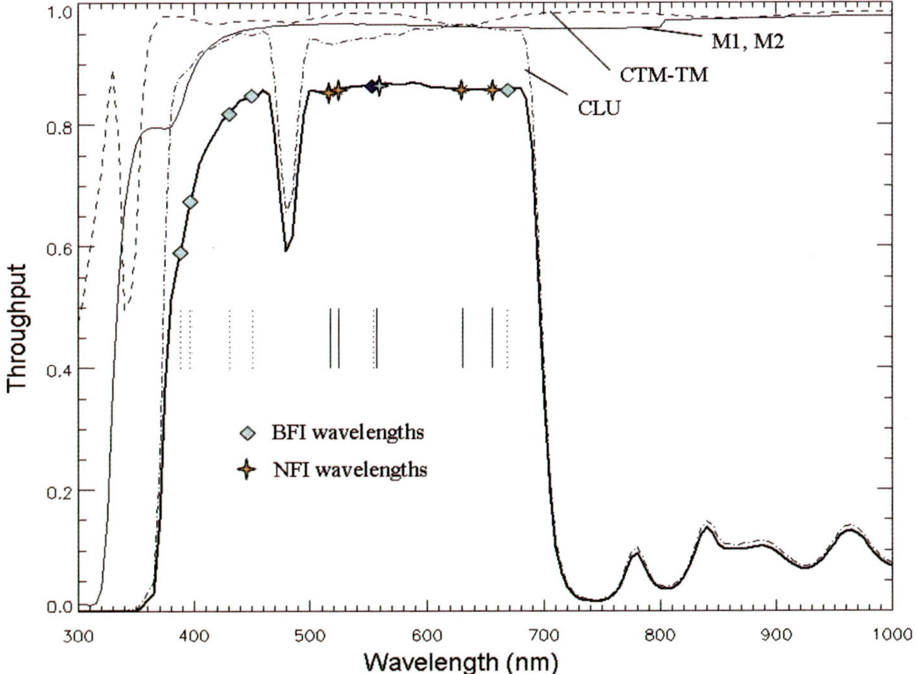

Figure 7 Reflectivity or transmissivity of each image forming the optical component and total throughput of the OTA. Symbols indicate the wavelengths used in the FPP. BFI: Broadband Filter Imager in the FPP; NFI: Narrowband Filter Imager in the FPP.

2. The sunshade and upper half of the shield tube work as a thermal radiator. The sunshade has an optical solar reflector facing the Sun to keep it cold; the upper area of the shield tube, not covered by multilayer insulation (MLI), is covered with a silverized Teflon sheet, a good IR radiator.

3. Solar heat (13 – 24 W) absorbed by the primary mirror is radiatively transmitted to the lower tube from its side and from a bottom cooling plate just beneath the mirror. The bottom cooling plate consists of a gold-plated aluminum honeycomb sandwich panel and radiatively absorbs the heat of the mirror from its back face. Decontamination heaters are attached to the back side of the bottom cooling plate.

4. Solar heat (≈ 1.5 W) absorbed by the secondary mirror is radiatively transmitted to the radiators from its back side.

5. Heat (≈ 2 W) absorbed by the 2FS, CLU, PMU, and CTM-TM and generated by their electronic components is conductively transferred to the mirror cell, emitted out through their housings, and is finally radiatively transmitted to the lower tube.

6. Heat (10 – 20 W) of the HDM is conductively transferred to the cylindrical structure supporting the HDM and outer spiders connecting the ring plate, and then radiatively transferred to the shield tube, the radiator, and space through the heat dump window.

7. The heat of the lower tube and the shield tube is radiatively emitted directly to the 3 K temperature of space through the entrance pupil and indirectly via the radiator of the sunshade and upper shield tube.

8. The OTA is thermally insulated from the spacecraft; The OTA is physically connected to the OBU only by the three mounting legs of low thermally conductive titanium and

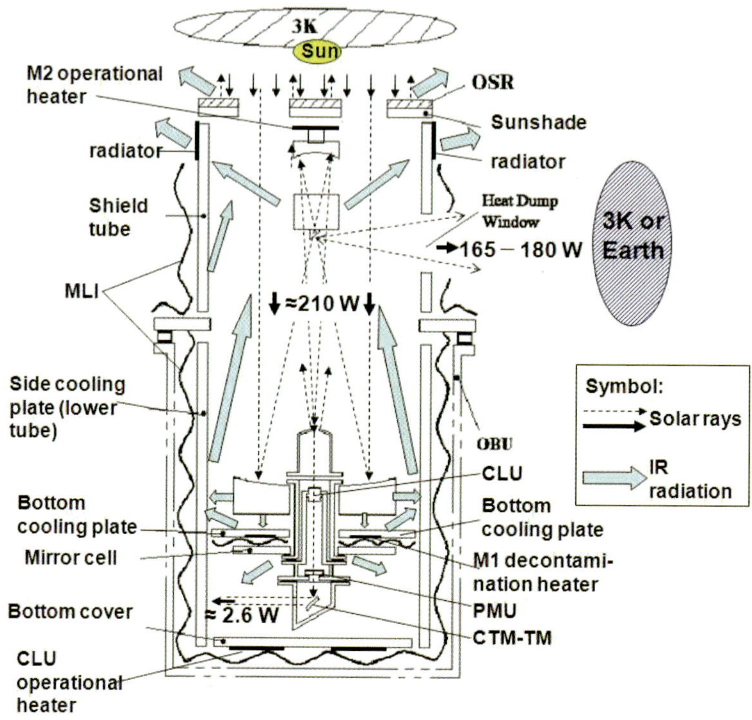

Figure 8 Schematic view of OTA thermal design and solar heat flow.

is radiatively decoupled from the OBU by MLI covering the lower tube and the bottom cover.

9. In the cold phase (*e.g.*, solar limb observations), the secondary mirror and CLU are warmed and maintained at their operational temperatures by dedicated operational heaters. The heater for the secondary mirror is attached to a separate heater plate behind the mirror, and for the CLU it is attached to the bottom cover. They warm up the optical elements indirectly by radiation to prevent localized temperature gradients and temperature ripples of optical elements resulting from the heater duty cycles.

10. In the postlaunch coldest phase, the critical optics are protected with survival heaters and later with decontamination heaters so that the temperature of the optics can be maintained about 10°C higher than their surrounding structures. The top and side doors, the coldest surfaces inside the OTA, are especially designed to work as cold plates, absorbing outgassing contaminants. Note that the side door should open before the top door to allow outgassing contaminants to escape away from the heat dump window.

During the thermal vacuum tests of the spacecraft thermal test model and flight model as well as optothermal tests of the OTA later described in Section 3.2, we confirmed that the thermal design of the OTA worked very well. We confirmed that the design is capable of maintaining the critical optical components and structures in their operational temperatures within large margins and also with small orbital variation, as well as achieving decontamination temperature. These tests were useful to refine the OTA thermal-math model predicting the temperature distributions on orbit.

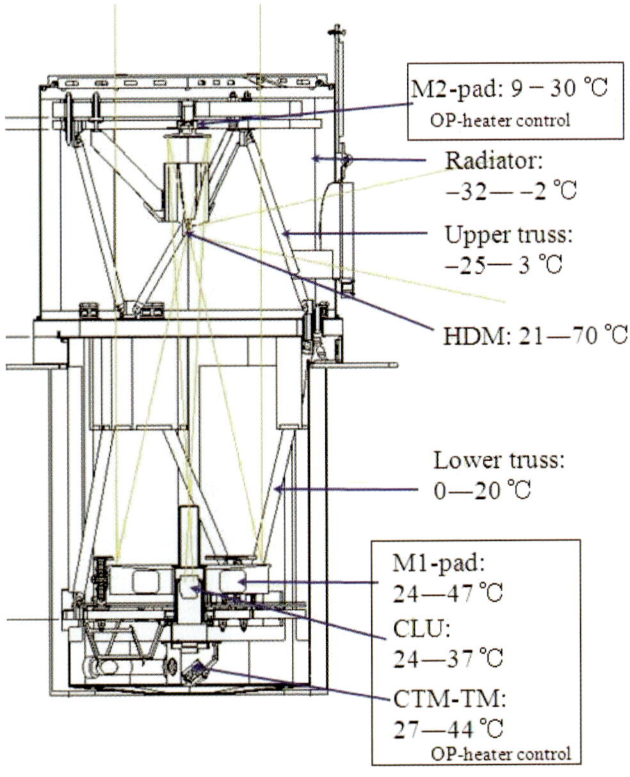

Figure 9 Predicted temperature distribution of on-orbit OTA for the cold case (at the beginning of life) and the hot case (at the end of life) of solar observations. In the cold case, M2 and CLU are maintained at their operational temperature by dedicated heaters.

For natural sunlight tests of the SOT, sunlight from the 54-cm-diameter beam was fed to the OTA entrance for an end-to-end optical check and rehearsal of actual solar observations. Unwanted sunlight rejection by the field stops (HDM and 2FS) was also confirmed (Figure 10).

The predicted temperatures of the on-orbit OTA are given in Figure 9 for the cold and hot solar observation cases.

It should be noted that the *Hinode* Sun-synchronous orbit gives day – night cycles for four successive months in a year, in which the OTA is exposed to a drastic thermal change. We identified this period as a degraded observation or nonobservation period and do not require full performance of the telescope.

3. Optics Alignment and Ground Testing

The OTA optical components were assembled in a dedicated tower structure having a dummy OBU and capable of rotating the OTA upside down. After the truss was installed on the dummy OBU, M1 was installed on the mirror cell and its optical axis was aligned in a vertical line, using a reticule (cross-hairs) set at the center of curvature of the M1 vertex radius. For this purpose, a reference theodolite was placed on the base of the tower, with its optical axis aligned with the center of mirror cell and the vertical line. Next, the ring plate assembled with the M2 and HDM units was installed on the top ring. Decenter, tilt, and despace of M2 with respect to M1 were adjusted by using a Shack – Hartmann sensor

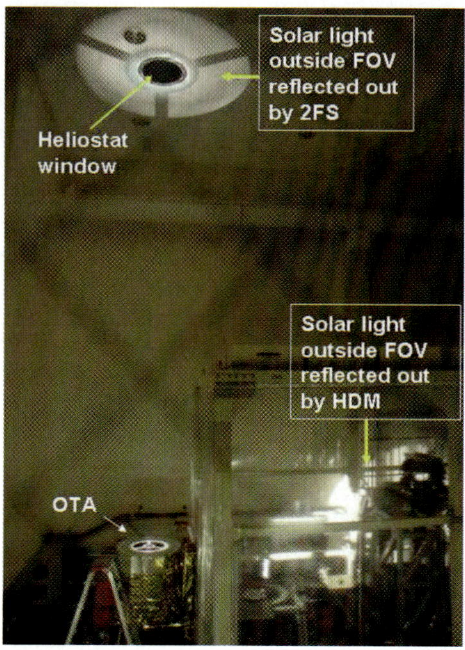

Figure 10 The natural sunlight ground test of the SOT. Sunlight of 54-cm beam diameter was fed to the OTA by a 90-cm heliostat installed on a cleanroom roof. This was an engineering first light of the SOT on the ground and allowed performance of end-to-end testing of the entire optics from the OTA through the CCD camera of the FPP and a rehearsal of actual solar observations expected on orbit. Images taken were not as good as those on orbit because of the severe gravity deformation of the optics and seeing effects as well. Unwanted sunlight rejection by the field stops (HDM and 2FS) of the OTA was also confirmed.

attached to the mirror cell. The sensor yielded real-time coefficients of low-order Zernike polynomials (*e.g.* Wilson, 1996) such as focus (A20), primary astigmatism (A22, B22), coma (A31, B31), and spherical (A40) aberration. A high-precision folding flat mirror (having a wavefront of 15 nm rms with a clear aperture of 600 mm) was installed at the top of the tower to face the OTA entrance pupil. The Shack–Hartmann sensor was focused at the center of the secondary field stop, and the coma aberration was eliminated by adjusting the decenter and tilt of the ring plate. At the same time, defocus was optimized by adjusting the despace of the ring plate, with allowance made for shrinkage of the CFRP truss owing to on-orbit dehydration.

3.1. Wavefront Error Measurement in the Zero-Gravity Condition

We measured the wavefront error using an interferometer after integrating the CLU, PMU, CTM-TM and ACL to the OTA (+1 G condition). The interferometer sent a collimated beam into the OTA from the exit pupil. With the folding flat above the entrance pupil, the double-path wavefront error of the OTA was obtained (Figure 11).

In the interferometer measurements, we employed a technique for deriving the phase (wavefront error) from a single interferogram (taken with a short exposure of 8 ms) using spatial heterodyning with high tilt of a reference flat. This method is less affected by the vibration of the test setup and the change in seeing conditions through a long optical path. In addition, the wavefront error of the interferometer itself was calibrated.

We carefully corrected phase unwrappings and phase gaps among three sections divided by the three spiders in restoring the phase from the interferogram. Typically, hundreds of usable interferograms, which were manually obtained, were used to achieve an accuracy of 5 nm rms for a single path. To calibrate a nonaxisymmetric wavefront error of the folding flat mirror (of confirmed small axisymmetric aberration in the flat mirror alone test), we

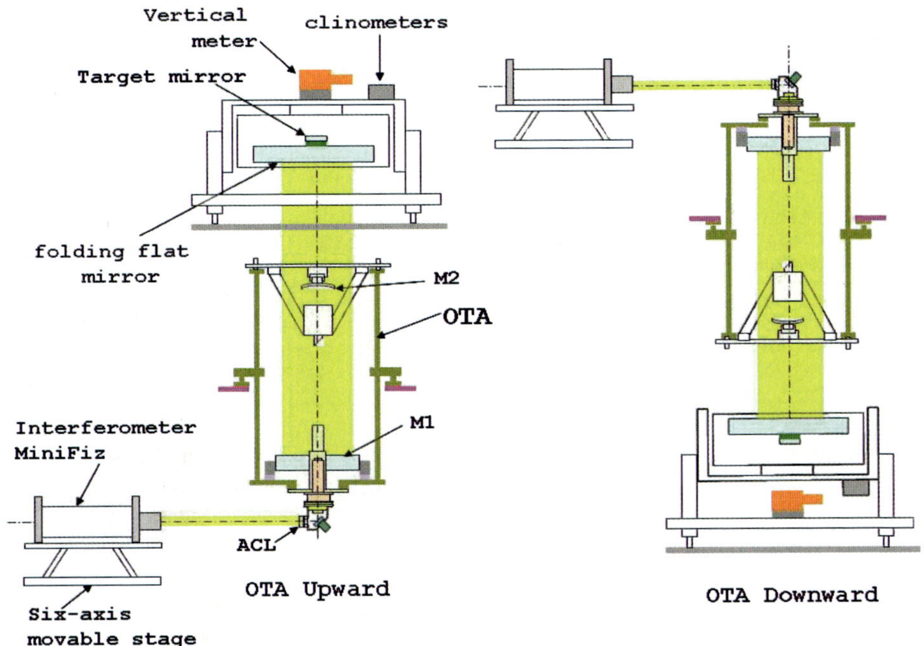

Figure 11 The setup configuration of zero-gravity WFE measurements for the OTA.

measured four sets of wavefront errors by rotating the flat by 90° steps around its axis and took their average. The result is given in Figure 12.

Finally, rotating the OTA upside down and setting the interferometer and the flat mirror for this configuration (Figure 11), we performed the measurements for a -1 G condition. By averaging the -1 G and the $+1$ G wavefronts, the effect of gravity can be canceled out and the OTA wavefront error in this zero-gravity condition was reduced. During these activities, mutual angles among the OTA pointing axis, exiting beam axis, and mechanical axis were measured by correlating the tilt of the folding flat mirror, image of the secondary field stop in the exiting beam, and the alignment cube, respectively.

Figure 12 shows the obtained wavefront error maps of the OTA. It is remarkable that, at the OTA in either the $+1$ G or -1 G condition, although a trefoil coma (coefficient B33) is a dominant aberration owing to a gravity deformation of the primary mirror, the trefoil coma was canceled in the zero-gravity condition and a superb wavefront map of OTA was obtained (18.2 nm rms for a single path).

The wavefronts of far-off-center field positions (about 132 arcsec radius) were also measured by tilting both the interferometer and the folding flat mirror. The results for four off-center positions are summarized in Table 3. Dominant aberrations of the off-center field are the primary astigmatism and defocus resulting from field curvature, as expected from the Gregorian design. However, their wavefront errors are small enough to give diffraction-limited performance; the average is 21.5 nm rms and the worst is 22.8 nm rms.

3.2. Optothermal Test

The thermal model predicts that the OTA on orbit will inevitably have a large temperature gradient along the optical axis; the bottom part of telescope including the primary mirror

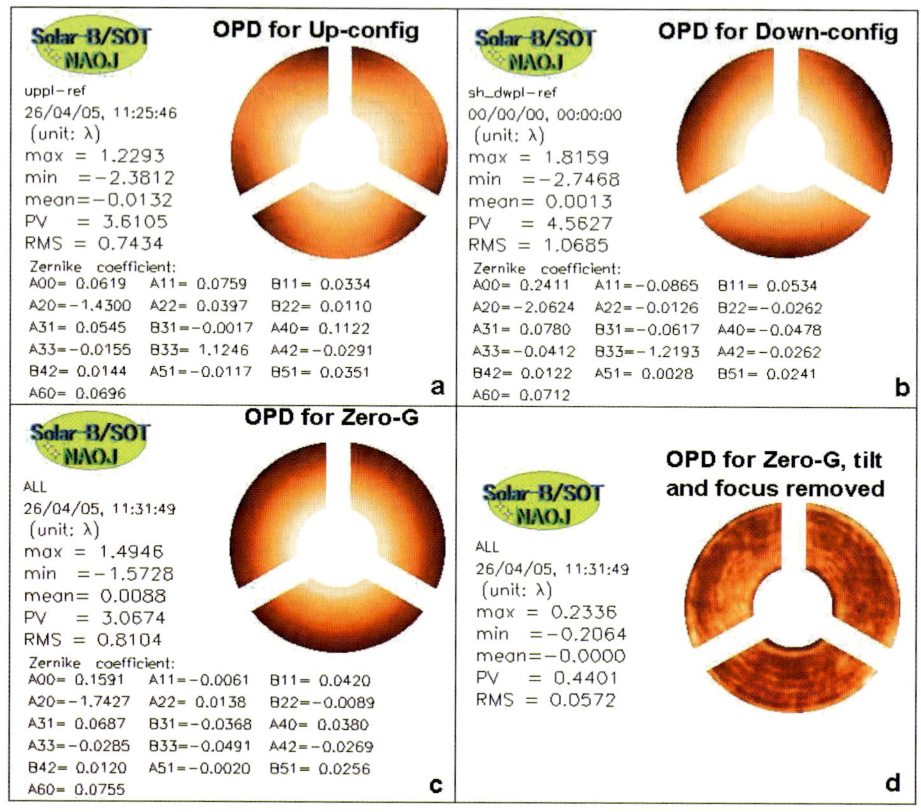

Figure 12 The wavefront error map of the OTA for the zero-gravity condition. The numbers in the figure are related to the wavefront and are in units of waves (with the test wavelength of He–Ne 632.8 nm) and for double-path measurements. The best-fit coefficients (Amn, Bmn) of Zernike polynomials are shown to give a major aberration mode: the focus (A20), primary astigmatism (A22, B22), coma (A31, B31), spherical (A40), trefoil coma (A33, B33), and so on. (a) The WFE for the OTA upward-pointing configuration ($+1$ G condition), (b) the WFE for the OTA downward-pointing configuration (-1 G), (c) the WFE averaged upward and downward for the zero-gravity condition, and (d) the average WFE-removed tilt and focus terms.

Table 3 Results of WFE measurements at various field positions. The positions are represented by incident angles (arcsec) in east–west and north–south direction on the Sun. In the measurements, the OTA pointed only upward and the wavefronts at the various field positions (bottom row) were evaluated by the subtraction of the field-center wavefront measurements. Then, the zero-gravity wavefront errors (middle row) were calculated with a root square sum of the field-center WFE of zero gravity and the difference WFE.

FOV (EW, NS)	$(0'', 0'')$	$(118'', 60'')$	$(118'', -60'')$	$(-118'', 60'')$	$(-118'', 60'')$
WFE (nm rms)	18.2	(22.8)	(21.2)	(21.2)	(20.8)
Difference from field-center WFE (nm rms)	–	13.7	10.8	10.9	10.1

will reach a temperature of $+24°C$ to $+47°C$, whereas the upper structures supporting the secondary mirror will reach a temperature of $9°C$ to $30°C$ depending on thermal conditions and its life phase. It is absolutely necessary, therefore, to verify the optical performance in

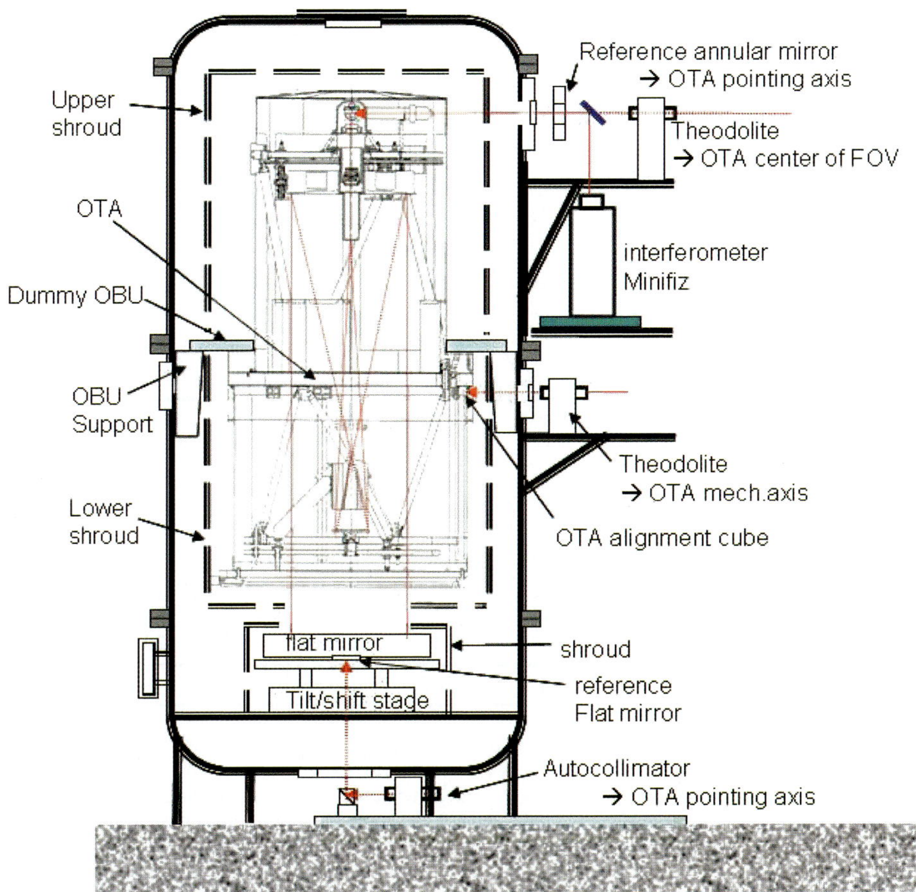

Figure 13 The OTA optothermal test configuration. The OTA was mounted in a vacuum chamber in an upside-down configuration with a dummy OBU made of superinvar, and the folding flat mirror with a tilt and rotational stage was installed at the base of the chamber. The interferometer mounted at the side wall fed the light into the OTA exit pupil through an optical-quality window.

such a temperature distribution even though the optomechanical design predicts that those temperatures are acceptable.

We conducted optothermal testing of the OTA using a dedicated vacuum chamber constructed for this purpose. The configuration of the test setup is shown in Figure 13. The OTA was mounted in a vacuum chamber in an upside-down configuration with a dummy OBU made of superinvar, and the folding flat mirror with a tilt and rotational stage was installed at the base of the chamber. Mounted on the side wall of the chamber, the interferometer looks into the OTA through an optical-quality window. The vacuum chamber has upper and lower shrouds that surround the entire OTA. By circulating temperature-controlled liquid through the shrouds, we changed the temperatures of upper and lower parts of the OTA independently over a range of $-40°C$ to $+40°C$. The temperature of the folding flat mirror was always kept about $20°C$, using another temperature-controlled shroud.

A sample test sequence is shown in Figure 14, together with the resulting WFE measurements for each test mode. Starting with room-temperature (hereafter referred to as room-T;

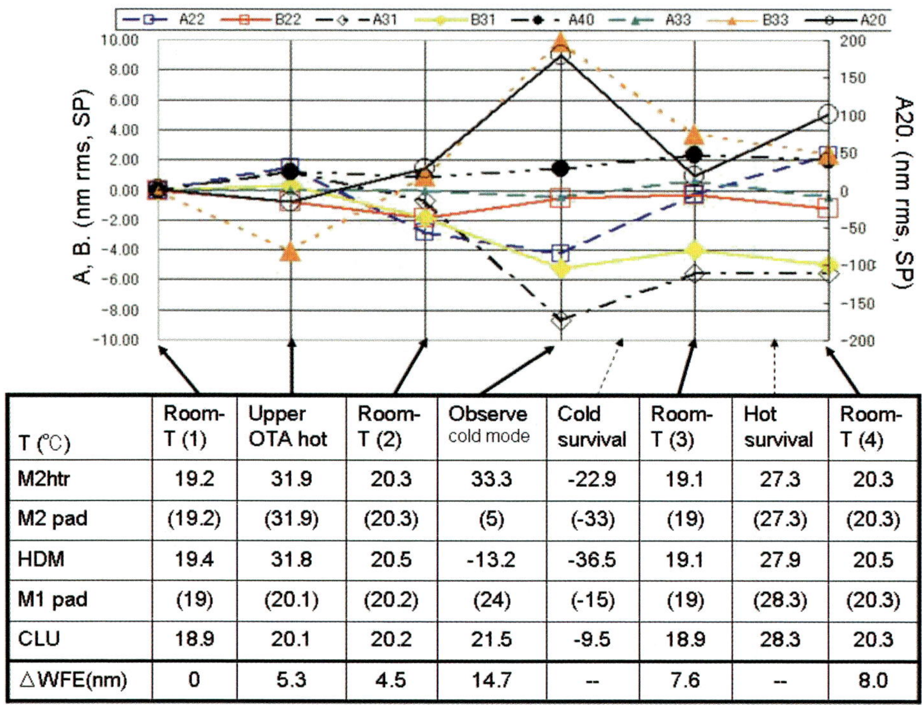

T (°C)	Room-T (1)	Upper OTA hot	Room-T (2)	Observe cold mode	Cold survival	Room-T (3)	Hot survival	Room-T (4)
M2htr	19.2	31.9	20.3	33.3	-22.9	19.1	27.3	20.3
M2 pad	(19.2)	(31.9)	(20.3)	(5)	(-33)	(19)	(27.3)	(20.3)
HDM	19.4	31.8	20.5	-13.2	-36.5	19.1	27.9	20.5
M1 pad	(19)	(20.1)	(20.2)	(24)	(-15)	(19)	(28.3)	(20.3)
CLU	18.9	20.1	20.2	21.5	-9.5	18.9	28.3	20.3
△WFE(nm)	0	5.3	4.5	14.7	--	7.6	--	8.0

Figure 14 Time sequence of one of optothermal tests for the flight model OTA. Starting with a room-temperature (room-T hereafter; temperatures of entire OTA are kept around 20°C) mode (1), the sequence went through upper-OTA hot mode, room-T mode (2), observation cold mode, cold survival mode, room-T mode (3), hot survival mode, and finally room-T mode (4). WFE was measured for each mode except for two survival modes. In the analysis, measured WFEs were subtracted from the WFE of the room-T mode (1) (the reference WFE) to clarify the effect of thermal change. The best-fit coefficients of low-order Zernike polynomials for obtained WFE maps are plotted for each test mode. Note that the ordinate on the right gives a scale of defocus coefficient (A20) and the ordinate on the left gives a scale of coefficient for other aberrations. Temperatures of some OTA components are given in a table. The temperatures in parentheses are predicted by using the OTA thermal-math model, because no temperature sensors are on the flight model mirrors. Differences of measured WFE from room-T mode (1) are given in the bottom row of the table.

temperatures of entire OTA are kept around 20°C) mode (1) giving a reference wavefront error, the sequence went as follows: the upper-OTA hot mode for check of defocus sensitivity, the room-T mode (2), the observation cold mode (solar limb observation in the cold phase), the cold survival mode for check of postlaunch coldest phase, the room-T mode (3), the hot survival mode for checking the entire OTA hot case owing to obliquely incident solar light into the OTA, and finally the room-T mode (4).

The WFE was measured for each mode except for the two survival modes. In the analysis, measured WFEs were subtracted from the WFE of the room-T mode (1) to clarify the effect of thermal change. The best-fit coefficients of low-order Zernike polynomials for obtained WFE maps are plotted for each test mode. Temperatures of some OTA components are given in a table in Figure 14. The temperatures in parentheses are predicted by using the OTA thermal-math model, because no temperature sensor is present for the flight model mirrors. Differences of measured WFE from that of the room-T mode (1) are given in the bottom row of the table.

First, notice the large focus change (A20) during the sequence. The main cause is dehydration shrinkage of the CFRP truss pipes connecting M1 with M2. The change amounted to $+100$ nm rms by the end of the test sequence, and from the trend the final shrinkage on orbit was evaluated to be about three times as large. Taking this shrinkage into account, we set the initial defocus to a large negative value (see Figure 12).

In the upper-OTA hot mode, we observed a defocus of 31 nm rms per 12°C after calibration of the dehydration effect. From additional extended experiments of focus sensitivity, we found that M2 movement along the optical axis owing to warping of the ring plate spider can cause noticeable focus change. The M2 supporting spider can displace in either of two ways: away from M1 by thermal expansion of metal parts on the ring plate (such as metal mounts attached for temperature sensor bases or cable ties or the adjustment mechanism rods for M2 alignment) or toward M1 by expansion of the HDM cylinder and its mounting outer spiders connected to the ring plate. It is likely that these two causes compete in magnitude but the former effect exceeds the latter a bit, giving in total the aforementioned focus sensitivity. The OTA thermal-math model predicts that the temperature of the ring plate and the HDM unit may change by a few degrees Centigrade in one orbital revolution. Case studies reveal that the possible orbital change in focus is at most about 8 nm rms, which is within a focal depth and is acceptable.

The wavefront error in the observation cold mode is not so large and is acceptable for diffraction-limited performance: root square sums of 18.2 (zero gravity) and 14.7 (thermal effect) give 23.4 nm rms for field center and root square sums of 18.2, 13,7, and 14.7 give 27.1 nm rms for the worst off-center position. The major aberration from thermal change is the trefoil coma of M1 and M2. Combining the results from the upper-OTA hot mode with those from the observation cold mode, we can estimate the sensitivity $\Delta B33/\Delta T$ as 0.85 nm rms K^{-1} for M1 and -0.4 nm rms for M2, where ΔT is the temperature deviation from room temperature. In the observation hot mode predicted toward the end of the OTA life, wavefront error only from the trefoil coma amounts to 19 nm rms. Provided that 10 nm rms WFE contributed from other aberrations, we have 28 nm rms in total for the field center. The number is a bit larger than the goal (25.8 nm rms), but it still meets with the diffraction-limit condition (less than 36.5 nm rms) as long as the OTA alone is concerned.

The change in coma aberration (A31, B31) is noticeable through the sequence and seems to be related to the change in defocus. The coma can be caused by the decenter and/or tilt of M2 and might be due to uneven shrinkage of the CFRP pipes by dehydration. It should be stressed that the OTA is robust both structurally and optically against severe thermal cycling of the cold and hot survival conditions; the change in WFE from the beginning to the end of the test is only 8 nm rms owing to coma aberration hysteresis.

3.3. Microvibration Test

We examined OTA pointing axis fluctuation caused by microvibration of mirrors excited by mechanical disturbances on the spacecraft, as this is critical for the OTA achieving diffraction-limited performance. Possible sources of the disturbance are momentum wheels (MW), inertial reference units (IRU; gyroscope), and moving mechanisms in observation instruments. Since the FPP correlation tracker and CTM-TM were designed to be able to suppress image jitter in frequencies less than 14 Hz, disturbances at higher frequencies are critical. To verify the image stability of the OTA on the spacecraft, we conducted a microvibration transmissibility test (Ichimoto *et al.*, 2004).

High-sensitivity accelerometers were attached to the primary and secondary mirrors of a high-fidelity mechanical test model to detect their displacement and tilt around two axes of x and y. The entire spacecraft was hanged from the cleanroom ceiling by four springs to avoid environmental disturbances. By operating the disturbing components or dummy shakers, the shift and tilt of mirrors were evaluated. In the test, we found that a resonance between the secondary mirror unit and one of the IRUs produces unacceptable pointing error. To improve upon this situation, therefore, we decided to change the operation frequency and the mounting location of the IRU.

With the flight model, we confirmed that the pointing jitter caused by microvibrations from MW, IRU, and most observation instruments were small enough to achieve the diffraction limit, being less than 0.03 arcsec rms. We identified a few movable elements, which are not frequently used for observation, that cause unacceptable pointing jitters and hence can only be operated during nonobservation time periods.

3.4. First-Light and On-Orbit Performance

The OTA optical performance was repeatedly verified with several postenvironmental test interferometer measurements for the OTA alone and the satellite until just before launch. The wavefront error of OTA has not changed within the measurement error of several nanometers rms, when defocus error is removed. We observed monotonic change in focus over months, which can be explained with an expansion of CFRP truss pipes connecting M1 with M2 by moisture absorption in air. This expansion will be mostly canceled by dehydration shrinkage of pipes on orbit.

Following the successful launch of the satellite on 23 September 2006 (JST), the SOT saw first light by the deployment of the OTA top door on 25 October 2006. Note that the side door had been opened eleven days before with decontamination heaters enabled for M1, M2, and the HDM so that OTA structure outgassing, a source of contaminants, could occur through the heat dump window.

The first-light images were taken with the FPP G-band (430-nm) filtergraph during the door deployment. Then a focus scan was performed at the disk center of the Sun to have the best focus position of the focusing (reimaging) lens, after the SOT temperatures stabilized. The OTA temperatures were about 10°C higher as a whole than predicted. Provided the temperature at the M1 and M2 pads are 10°C higher than that predicted for the cold case given in Figure 9, and any aberration additional to trefoil coma is 10 nm rms, from the discussion in Section 3.2, we can predict that the WFE of the OTA would be of 16 nm rms excess owing to thermal change on orbit, and in total 24.2 nm rms and 28 nm rms at the field center and around 130 arcsec off-center, respectively.

An example of the G-band images in focus is shown in Figure 15. Note that the correlation tracker image-stabilization system was not yet used at this time. Nevertheless, the G-band images are superb, showing many point-like bright features of about 0.2 arcsec wide. To check the resolution, the intensity profile of the point-like bright features was compared with an ideal point spread function (PSF) of the OTA as shown in Figure 15. The profile of the bright point can be explained if its true profile is a two-dimensional Gaussian of 0.16 arcsec FWHM; the profile of the PSF convolved with the Gaussian gives the observed one. The bright point of 0.16 arcsec width makes sense and this implies that the G-band filtergraph of the SOT has diffraction-limited performance. Therefore, there is no doubt that the OTA keeps the superb diffraction-limited performance on orbit.

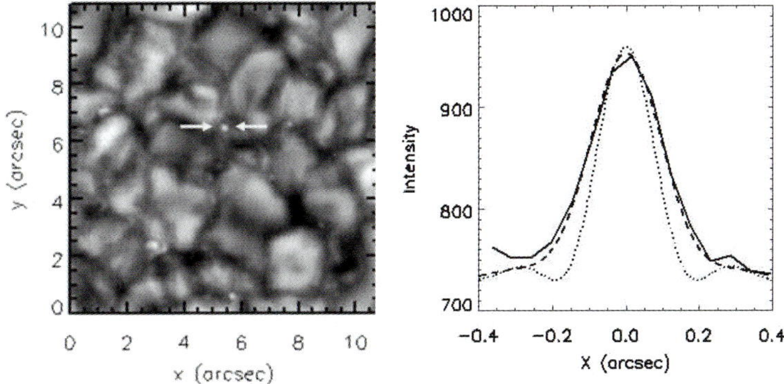

Figure 15   Image taken in the G band (430 nm) adjusted in focus after first light on orbit (left). The exposure is 30 ms and the scale of a CCD pixel is 0.054 arcsec. The solid curve represents the intensity profile of a bright point between two arrows in the image. The dotted curve is a profile of the ideal PSF of the OTA and the dashed curve is a profile of PSF convolved with a two-dimensional Gaussian of 0.16 arcsec FWHM.

4. Conclusion

We have described in detail the instrument design and ground testing of the Optical Telescope Assembly of the Solar Optical Telescope aboard the *Hinode* satellite. In short, the OTA consists of the aplanatic Gregorian with the well-designed collimating lens unit near the center of the primary mirror, followed by a polarization modulator unit, an active tip-tilt mirror, and an astigmatism corrector lens.

Owing to the strict budget control of wavefront errors of each optical component and the elaborate optical tests of components and of the entire OTA simulating the on-orbit condition, we are confident that the OTA can achieve diffraction-limited seeing on orbit. The definition of diffraction limit here is a wavefront error of less than 36.5 nm rms (Strehl ratio larger than 0.8 at 500 nm), and the goal of the OTA was 25.8 nm rms WFE in the field center. It is expected that the OTA wavefront error is about 24 nm rms around the field center and 28 nm rms near the field of 132 arcsec radius as long as the primary mirror is not significantly contaminated. Should solar light absorptance of the mirror increase by 5%, this would increase the WFE to about 30 nm rms in the edge of this field area. The focus is stable; the change in focus around an orbit is expected to be negligibly small, at most 8 nm rms.

In addition, the OTA is capable of high photon throughput in the observation wavelengths from 388 to 668 nm, diffraction-limited imaging in the FOV of 361 × 197 arcsec, limited by the secondary field stop at the Gregorian focus, and off-limb observations up to 200 arcsec (field center), limited by the diameter of the heat dump mirror at the primary focus.

As the first-light images from the SOT have demonstrated, it has already begun to provide unprecedented, continuous solar optical data of high spatial resolution. We conclude that the OTA is the largest state-of-the-art solar telescope that has ever been completed and flown in space.

Acknowledgements   The SOT aboard *Hinode* was developed by a joint collaboration of Japanese and U.S. instrument teams, and the authors would like to thank all the members of the teams. Particularly, they thank ISAS/JAXA; Professors T. Kosugi (the late general manager of *Solar-B*), K. Minesugi (structural), and A. Ohnishi (thermal), MELCO team; Messrs. Y. Hasuyama (integration, quality control, and test), N. Yoshida, O. Takahara (guidance control and microvibration), T. Ozaki (composite material), N. Kaido (OTA thermal), and J. Nakagawa (door deployment) and the FPP team of LMSAL; Drs. A. Title, T. Tarbell, and W. Rosenberg.

They are also deeply obliged to SAGEM/REOSC; Messers. R. Mercier-Ythier, L. Thepaut, E. Ruch, and D. Mouricaud for their tremendous contribution in developing the OTA mirrors, its integration procedure, and space-qualified silver coatings. The CLU was manufactured by Canon and they are grateful to Messrs. H. Yokota and M. Suzuki for their superb work. The superior ACL was fabricated by Sankyo Kogaku Industry and AR-coated by Okamoto Optics Works. They thank Mr. S. Abe (formerly of Canon) and Ms. Y. Sakakibara of Genesia Corp. for their contribution in development and maintenance of the CLU and ACL and Mr. K. Waseda of NAOJ for his superb work on deposition of the IR/UV rejection multilayer coating of the CLU. Last but not least, they thank Prof. T. Sakurai, Drs. Y. Hanaoka and R. Kano of NAOJ, Dr. M. Akioka of NiCT, and Dr. M. Nishio (formerly of NAOJ) of Kagashima University for their contribution in the OTA initial designing phase and Mr. Miyashita of NAOJ, Dr. S. Nagata of Kyoto University, and Drs. Y. Sakamoto and N. Kohara of the University of Tokyo for their assistance in the OTA and spacecraft-level testing. Finally, they would like to express their deep appreciation to Mr. T. Cruz of LMSAL for his elaborate review in revising the English of this manuscript.

References

Culhane, J.L., Harra, L.K., James, A.M., Al-Janabi, K., Bradley, L.J., Chaudry, R.A., *et al.*: 2007, *Solar Phys.* **243**, 19.

Dunn, R.B.: 1981, *Space Sci. Rev.* **29**, 341.

Guimond, S., Elmore, D.: 2004, *OE Mag.* **5**(5), 26.

Golub, L., DeLuca, E., Austin, G., Bookbinder, J., Caldwell, D., Cheimets, P., *et al.*: 2007, *Solar Phys.* **243**, 63.

Ichimoto, K., Tsuneta, S., Suematsu, Y., Shimizu, T., Otsubo, M., Kato, Y., *et al.*: 2004, In: Mather, J.C. (ed.) *Optical, Infrared, and Millimeter Space Telescopes, Proc. SPIE* **5487**, 1142.

Ichimoto, K., Lites, B., Elmore, D., Suematsu, Y., Tsuneta, S., Katsukawa, Y., *et al.*: 2008, *Solar Phys.* in press.

Kano, R., Sakao, T., Hara, H., Tsuneta, S., Matsuzaki, K., Kumagai, K. *et al.*: 2008, *Solar Phys.* in press.

Kosugi, T., Matsuzaki, K., Sakao, T., Shimizu, T., Sone, Y., Tachikawa, S., *et al.*: 2007, *Solar Phys.* **243**, 3.

Ozaki, T., Ikeda, C., Isoda, M., Tsuneta, S.: 1996, In: Rust, D.M. (ed.) *Missions to the Sun, Proc. SPIE* **2804**, 22.

Schroeder, D.J.: 2000, *Astronomical Optics*, 2nd edn., Academic Press, San Diego.

Shimizu, T., Nagata, S., Edwards, C., Tarbell, T., Kashiwagi, Y., Kodeki, K., *et al.*: 2004, In: Mather, J.C. (ed.) *Optical, Infrared, and Millimeter Space Telescopes, Proc. SPIE* **5487**, 1199.

Shimizu, T., Nagata, S., Tsuneta, S., Tarbell, T., Edwards, C., Shine, R., *et al.*: 2008, *Solar Phys.* in press.

Title, A.M.: 1989, In: von der Lühe, O. (ed.) *High Spatial Resolution Solar Observations, Proc. 10th Sacramento Peak Summer Workshop*, National Solar Observatory, 35.

Tarbell, T., *et al.*: 2008, *Solar Phys.* to be submitted.

Tsuneta, S., Ichimoto, K., Katsukawa, Y., Nagata, S., Otsubo, M., Shimizu, T., *et al.*: 2008, *Solar Phys.* submitted.

Wilson, R.N.: 1996, *Reflecting Telescope Optics I*, Springer, Berlin.

Image Stabilization System for *Hinode* (Solar-B) Solar Optical Telescope

T. Shimizu · S. Nagata · S. Tsuneta · T. Tarbell · C. Edwards · R. Shine ·
C. Hoffmann · E. Thomas · S. Sour · R. Rehse · O. Ito · Y. Kashiwagi · M. Tabata ·
K. Kodeki · M. Nagase · K. Matsuzaki · K. Kobayashi · K. Ichimoto · Y. Suematsu

Originally published in the journal Solar Physics, Volume 249, No 2.
DOI: 10.1007/s11207-007-9053-z © Springer Science+Business Media B.V. 2007

Abstract The *Hinode* Solar Optical Telescope (SOT) is the first space-borne visible-light telescope that enables us to observe magnetic-field dynamics in the solar lower atmosphere with $0.2 - 0.3$ arcsec spatial resolution under extremely stable (seeing-free) conditions. To achieve precise measurements of the polarization with diffraction-limited images, stable pointing of the telescope (< 0.09 arcsec, 3σ) is required for solar images exposed on the focal plane CCD detectors. SOT has an image stabilization system that uses image displacements calculated from correlation tracking of solar granules to control a piezo-driven tip-tilt mirror. The system minimizes the motions of images for frequencies lower than 14 Hz while the satellite and telescope structural design damps microvibration in higher frequency ranges. It has been confirmed from the data taken on orbit that the remaining jitter is less than 0.03 arcsec (3σ) on the Sun. This excellent performance makes a major contribution to successful precise polarimetric measurements with $0.2 - 0.3$ arcsec resolution.

K. Kobayashi now at NASA/Marshall Space Flight Center, Huntsville, AL 35812, USA.

T. Shimizu (✉) · K. Matsuzaki
Institute of Space and Astronautical Science, Japan Aerospace Exploration Agency,
3-1-1 Yoshinodai, Sagamihara, Kanagawa 229-8510, Japan
e-mail: shimizu.toshifumi@isas.jaxa.jp

S. Nagata
Hida and Kwasan Observatories, Kyoto University, Kamitakara, Gifu 506-1314, Japan

S. Tsuneta · K. Kobayashi · K. Ichimoto · Y. Suematsu
National Astronomical Observatory of Japan, Mitaka, Tokyo 181-8588, Japan

T. Tarbell · C. Edwards · R. Shine · C. Hoffmann · E. Thomas · S. Sour · R. Rehse
Lockheed-Martin Solar and Astrophysics Laboratory, Palo Alto, CA 94304, USA

O. Ito · Y. Kashiwagi · M. Tabata · K. Kodeki
Mitsubishi Electric Corp., Amagasaki, Hyogo 661-8661, Japan

M. Nagase
Systems Engineering Consultants Corp., Shibuya, Tokyo 150-0031, Japan

Keywords Space vehicles · Telescopes · Instrumentation: adaptive optics · Correlation tracker · Image stabilizer · Tip-tilt mirror · Sun: photosphere · Magnetic fields · Chromosphere

1. Introduction

Of the principal scientific goals of the *Hinode* (formerly Solar-B) mission (Kosugi *et al.*, 2007), precisely measuring the properties of magnetic fields and their dynamics at the photospheric and chromospheric layers is the prime observational requirement for the Solar Optical Telescope (SOT; Shimizu, 2005; Tsuneta *et al.*, 2007). Magnetic fields at the photosphere are extremely fragmented with fine structures on scales that are especially difficult to resolve from the ground where seeing effects degrade spatial resolution. The SOT is the largest aperture (50 cm in diameter) and most advanced space telescope dedicated to solar observations in visible-light wavelengths. After the successful launch of the *Hinode* spacecraft, SOT has successfully begun observations from November 2006 to provide a continuous series of diffraction-limited images (0.2 – 0.3 arcsec at 388 – 668 nm) and precise measurements of polarization. The image stabilization system installed in SOT, which is described in this paper, makes a major contribution to realizing SOT's superior performance for spatial resolution and polarization measurements. The aim of the image stabilization system is to remove the motion of the solar images on the focal plane of SOT.

Image stabilization systems have been developed at ground-based observatories in the past couple of decades to remove jitters in solar images on the focal plane of the telescope that are due to atmospheric seeing effects and mechanical vibrations caused by wind shaking. Starting from spot trackers, which track a pore, sunspot, or other high-contrast features on the solar disk, granulation correlation trackers have recently become the standard image stabilizer for ground-based observations. These measure motions based on the solar granulation pattern at the region of interest. Until now, only a limited number of granulation correlation trackers have been stably operated at some ground-based optical observatories (van der Lühe *et al.*, 1989; Ballesteros *et al.*, 1996; Molodij *et al.*, 1996), giving significant improvements to observational capabilities on the ground. The image stabilization system of SOT is the first successful application of a correlation tracker for a space-based instrument. Other recent space-based instruments with high spatial resolution capabilities have used an error signal from a limb sensor or positional sensor on the focal plane of a guide telescope to remove jitter. The *Transition Region And Coronal Explorer* (TRACE) satellite has an image stabilization system that provides jitter removal better than 0.1 arcsec rms based on an error signal from a guide telescope attached beside the main telescope (Handy *et al.*, 1999). This kind of system is an open-loop control system in the sense that the error signal from the guide telescope is used to control the tilt of the secondary mirror of the main telescope without feeding back image motions measured on the focal plane of the main telescope.

To achieve diffraction-limited resolution for precise measurements of polarization with polarimetric accuracy of about 0.1% or better (Ichimoto *et al.*, 2007), stable pointing better than 0.09 arcsec (3σ) is required for the SOT observations. This level of stabilized pointing during multiple exposures taken at different phases of the polarization modulator significantly reduces false signals in the combined (added and/or subtracted) data representing the polarization, such as the longitudinal magnetogram and Stokes $IQUV$ parameters. Note that the Spectro-Polarimeter, which is one of the focal plane instruments of SOT, also records two spectra simultaneously in orthogonal linear polarizations for further reduction of crosstalk between pointing errors and polarization. Because the degradation of the point spread function (Strehl ratio) can be well evaluated by using time profiles of the pointing jitter, the requirement is specified in 3σ rather than in 0-p. A requirement of 0.09 arcsec (3σ) corresponds to Strehl error of ≈ 0.946 at $\lambda = 500$ nm.

An extremely stable condition is achieved by a combination of an active image stabilization system and the structural-thermal design of the telescope and spacecraft. The image stabilization system can reduce image jitters for frequencies lower than about 20 Hz, whereas frequencies higher than about 20 Hz need to be damped by the structural design of the telescope and spacecraft. The solar pointing direction (Z axis) of the satellite body shows low-frequency jitters and slow drifts that can be caused by uncertainties in the satellite attitude (which is based on signals from the attitude-sensing devices such as sun sensors and gyroscopes) and any residual errors remaining after controlling the spacecraft body attitude. A gradual drift can be caused by the thermal deformation of the spacecraft mechanical structure on which the attitude-sensing sensors are mounted. In the frequency range higher than about 20 Hz, internal disturbances located inside the spacecraft may excite small vibrations in the spacecraft mechanical structures and the telescope optics. Significant efforts have been taken to control such high-frequency jitter (*i.e.*, microvibration): (1) the sources of microvibration on board, such as attitude control gyroscopes and telescopes' filter wheels, are evaluated with analytical estimates and measurements and (2) the mechanical structures of the telescope and spacecraft are designed to reduce the propagation of the microvibration into the telescope and the optical elements. Finally, the jitter level has been measured and evaluated on the spacecraft level as one of the major ground tests, using pointing error signals generated by the correlation tracker with a laser point source (Takahara *et al.*, 2004). For frequencies lower than about 20 Hz, 0.02 arcsec (3σ) was required as the residual error in the error budget allocation to ensure the overall 0.09 arcsec scientific requirement. The functions and performance of the image stabilization system were measured and evaluated in the laboratory environment before installing it into SOT (Shimizu *et al.*, 2004).

2. Image Stabilization System

The stabilization system consists of a correlation tracker and a piezo-driven tip-tilt mirror with servo control electronics. Figure 1 shows a schematic diagram describing the overall configuration of the system. The tip-tilt mirror mechanism (CTM-TM) is located in the collimated beam behind the primary mirror (Suematsu *et al.*, 2007) and folds the incident beam into the direction toward the focal plane package (FPP; Tarbell *et al.*, 2007). The correlation tracker (CT) is a high-speed CCD camera with a real-time correlation algorithm on the FPP computer (FPP-E) that produces error signals estimating the displacement of the granule patterns seen in the field of view. The CT camera is in the focal plane package (FPP). The beam splitter at the entrance of the FPP optical bench feeds a small amount of light into the CT optical path. A wedge wheel is placed in front of the CCD to provide an offset of the tracking area.

CT camera images are transferred in real time to the flight computer (FPP-E), where they are processed with a real-time correlation algorithm (Section 2.1). Receiving pointing error (displacement) signals from FPP-E, the servo control computer (CTM-E) controls the tilt angle for the tip-tilt mirror, reducing the CT residual signal (Section 2.3). The piezo electronics (CTM-TE) generates voltages to drive three piezo actuators in the tip-tilt mirror mechanism.

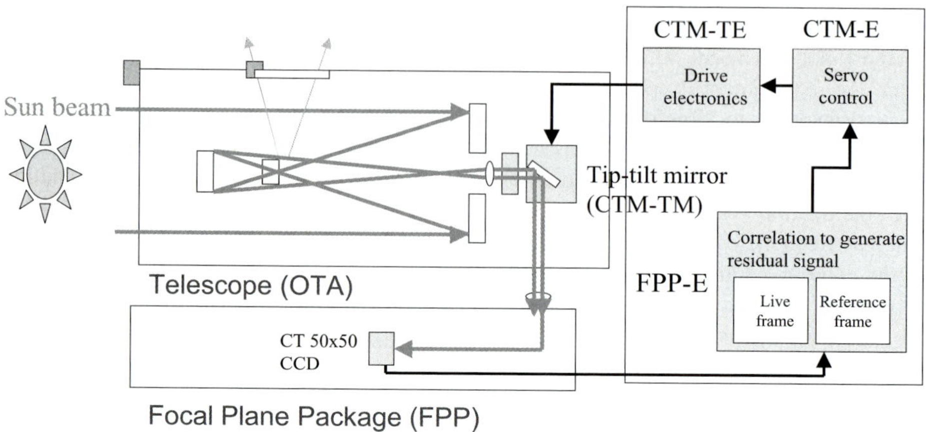

Figure 1 Image stabilization system overview.

2.1. Correlation Tracker

The CT CCD camera produces a continuous series of frames with 50×50 pixels at a frame rate of 580 Hz. The plate scale is 0.22 arcsec pixel^{-1}. The camera obtains images of solar features seen in visible wavelength (629 – 634 nm), such as granules and tiny pores. Since the typical size of granules is 1 – 2 arcsec in diameter, several granules are in the field of view. The series of camera live frames is processed in real time in the FPP-E to generate a pointing error signal representing the displacement of solar features in the live image with respect to a reference image. The correlation calculation uses pixel summations of the absolute difference between the live and reference image calculated for multiple shifts of the reference image by $\pm(1-3)$ pixel. The subpixel location of the minimum residual point is then determined by performing a two-dimensional polynomial fit to the sums. When image shifts are less than about 0.44 arcsec (2 pixels), the algorithm produces an error signal proportional to the image displacement on the CT CCD. A complicated algorithm is designed to take care of various situations, such as live frames with a large shift. Shifts up to about ± 5 pixels (± 1.1 arcsec) generate corrections but when the shift is more than 0.44 arcsec only the computed direction is generated with a saturated magnitude of 0.44 arcsec because the shift becomes less inaccurate. For image shifts larger than about 1.1 arcsec, error signals are not fed back to the tip-tilt mirror control. Gradients are not removed from the image before calculating the pixel summation of the absolute difference. Intensity gradients across the image can lead to slow drifts of the tracked image. When solar features near the limb are observed, an intensity gradient caused by limb darkening exists across the image. This gradient gives a small bias to the pointing error signal in the intensity gradient direction, resulting in a slow drift of the tracked image.

Solar granulation patterns on the photosphere evolve with time scales of several minutes. As the reference image becomes old, there is less correlation with the live image. Thus, the reference image is automatically updated every ≈ 40 seconds (adjustable by command). When the reference image is changed to a new image, the correlation may generate a small deviation on the order of 0.01 to 0.02 arcsec because the two different frames taken with 40 seconds separation as a reference, but the magnitude of this deviation is much smaller than the required jitter stability (0.09 arcsec).

The CT tracking region can be changed by inserting one of the wedge plates on the wedge wheel. The purpose of these wedge offsets is to ensure that the CT camera field of view has a bright, high-contrast granulation field without excessive foreshortening to track, regardless of whether SOT is pointing at the limb or at the center of a sunspot. The wedges offset the field of view by about 130 arcsec in the east–west direction and about 65 arcsec in the north–south direction. These offsets were chosen so that the CT field of view would always be in the field of view of the Narrow-band Filter Imager (NFI; 328×164 arcsec). The wedge offsets are routinely used when observing the limb or for active regions when the pointing is centered on a sunspot. We have found that the solar flows in sunspot moat regions are large enough to disturb long-term pointing at active regions, so the wedges are often used when pointing is anywhere near a large leading sunspot.

2.2. Tip-Tilt Mirror

The mirror, which is 60 mm in diameter and 10 mm thick, reflects the beam at an angle of 45 degrees. The mirror is tilted by a steering mechanism, which uses three piezoelectric translators (PZTs) placed 120 degrees apart from each other behind the mirror (Figure 2). PZTs are ceramic actuators that directly convert electrical energy into small linear motion. They extend ≈ 30 µm by applying ≈ 100 V and provide fine positioning by controlling the applied voltage. The tilt range over which the tip-tilt mirror can move is a hexagonal area defined by the triangle stations of the three PZTs in the mechanism. The flight-model tip-tilt mirror can be tilted over a full range of 314 arcsec around the X axis and 254 arcsec around the Y axis. Converting these to the sky angle gives a the maximum allowable tilt angle of ± 13.3 arcsec in the solar north–south direction and ± 15.4 arcsec in the east–west direction. The mirror tilt range covers 10.5 arcsec in any direction from the center position.

2.3. Servo Controller

The pointing error signal generated by the correlation tracker is sent to the servo controller (CTM-E) via a digital line. The closed-loop control is implemented in the software of this controller electronics. Because the main purpose of this system is to suppress the low-frequency disturbance, the software basically works as an integration controller. The pointing error signal used for the closed-loop control has a relatively large delay time of 3.2 ms. The major contribution comes from the image exposure and readout of the pixels from the CCD (2.6 ms). The other contributions are image correlation analysis, signal transfer to the servo controller, and servo process in the servo controller. The choice of the crossover frequency for this stabilization system depends on the disturbance environment on the spacecraft. The crossover frequency can be adjusted in the 10–30 Hz range by command after evaluating on-orbit performance. Before flight, we chose 14 Hz as the nominal initial value for the crossover frequency. In general, a higher crossover frequency is suitable for stably suppressing jitters in the control band below the frequency. But the damping performance becomes worse in the 30–80 Hz range with higher crossover frequency and the jitter in this range becomes amplified by the stabilization system. The operating frequency of the momentum wheels, which are operated as actuators of the satellite attitude control in the spacecraft, are located around 30–40 Hz. If a high frequency is chosen for the crossover frequency, the microvibrations in 30–40 Hz caused by the momentum wheels may be amplified by the stabilization system, resulting in larger jitters on the focal plane of SOT. With a crossover frequency of 14.0 Hz, the system has a phase margin of 69.2 degrees and a gain margin of 13.2 dB, which means that the servo system is very stable.

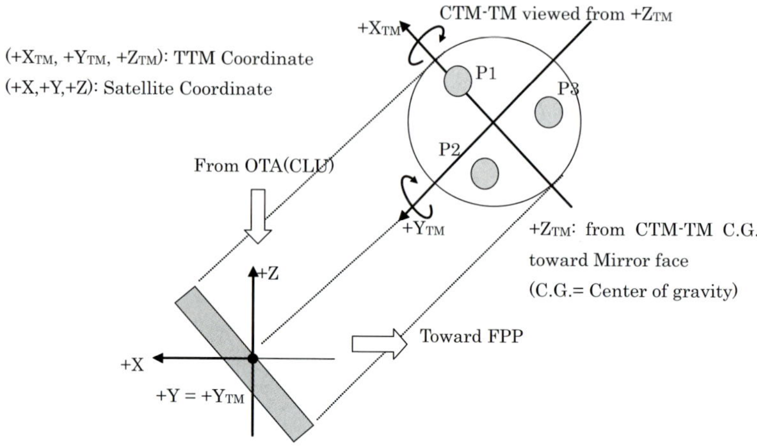

Figure 2 Tip-tilt mirror schematic configuration.

The control signal is transformed to a voltage command to the three PZTs. In addition to the closed-loop control with feedback of the pointing error signal, the controller has another mode for tilting the mirror by a tilt angle command (open-loop control). Details of the servo control can be found in Shimizu *et al.* (2004).

3. Coordinated Operations with Observations

3.1. Target Tracking

The correlation tracker produces displacement of the granules seen in the 11×11 arcsec field of view, and therefore the tip-tilt mirror is controlled to track the group motion of the granules located in the small area. If, for example, the granules in the small area show a group motion of 0.3 km s^{-1}, the entire SOT field of view slowly drifts by 1.5 arcsec in an hour. This drift can be removed in the final series of SOT images by applying image correlation to a much larger field of view when the data are analyzed.

When the telescope's pointing is directed to a fixed heliocentric position on the solar disk (fixed pointing mode of the spacecraft attitude control), the region of interest moves toward the solar west at the rate of solar rotation (≈ 0.15 arcsec min^{-1}). Since the stroke range of the tip-tilt mirror is too narrow to cover this motion for longer than about 60 min, the spacecraft attitude is primarily controlled to follow an observing region at the rate of the solar differential rotation. Thus, the tip-tilt mirror can continuously work for a fairly long time (a few hours or so) without reaching the stroke limit.

When the tip-tilt mirror mechanism hits the stroke limit, the onboard computer automatically resets the tilt angle to its home position and resumes the servo control from the home position. This means a sudden positional jump occurs in the series of images at the time of tip-tilt angle reset. This positional jump needs to be removed by the data processing on the ground, by using the image correlation of the images before and after the jump. Also, the tip-tilt mirror angle reset may be executed at the start of a mapping by the Spectro-Polarimeter, to reduce the possibility that positional jumps caused by the tip-tilt mirror reset are included in long slit-scanned Stokes map data.

3.2. Mirror Tilt Angle

When the tip-tilt mirror angle is reset to the home position, it may not be located exactly at the center of the stroke. Typically, the home position may have about a 1 arcsec uncertainty on the sky angle owing to the memory effect of PZT hysteresis. Figure 3a shows the home positions returned from eight extreme positions (10 arcsec off) after a reset of the tip-tilt mirror mechanism. The mechanism has a hysteresis of about 17% of the commanded motion as shown in Figure 3b. For the reset, open-loop control is used; this is accomplished by simply applying voltages for the nominal home position. It should be noted that hysteresis is completely eliminated during closed-loop control for positioning, in which the tilt angle is controlled by a servo loop with pointing errors from the correlation tracker.

The tip-tilt mirror mechanism does not have any sensors for measuring the actual absolute tilt angle in the mechanism. The only angle information available is the angle commanded to the tip-tilt mirror drivers. As shown in Figure 3, the commanded angle is not equal to the actual tilt angle owing to the hysteresis. The scaling of the angle commanded to the tip-tilt mirror mechanism relative to the actual angle was determined from the measurement over the full voltage range (-15 to 80 V). When a series of small (subarcsec) angles are applied to adjust the tip-tilt mirror angle, which is the case for most periods in servo (closed-loop) control, the magnitude of the actual tilt angle change is about half of the commanded angle change. In servo control mode, the feedback ensures that the tip-tilt mirror is controlled to keep the position of solar features in the field of view unchanged. In both open- and closed-loop modes, the angles commanded to the actuators are provided in telemetry and in image headers. For large angle changes commanded, the angle information in the telemetry is almost equal to the actual tilt angle. For small angle changes commanded, the actual tilt angle is significantly smaller than the angle in the telemetry. If the angle information is used in scientific data analysis, users should understand this hysteresis effect.

4. System Performance

The performance of the flight-model image stabilization system after combining the correlation tracker and the tip-tilt mirror was evaluated in the laboratory environment. This evaluation proved that the performance of the flight system is excellent and meets all the requirements. The test setup had ambient jitter of $0.03-0.09$ arcsec (3σ), depending on various configurations used in the test. The flight system was tested on a floating optical bench in a normal laboratory vibration environment. The major contribution to the ambient jitter came from several numbers of resonance peaks observed in the $40-150$ Hz range, frequencies higher than the crossover frequency of the stabilization system, that were excited by external disturbances. We identified some of them as the resonance frequencies of the mechanical test supports used on the floating optical bench for holding optical elements and the flight model. When the servo was closed, the low-frequency disturbance was dramatically reduced, as expected from the jitter reduction factor. Figure 4 shows the ratio of the square root of the power spectral density of that random jitter with the servo off over the power spectral density with the servo on. The reduction factor is ≈ 50 at low frequencies, ≈ 10 at 1 Hz, and ≈ 1 at 15 Hz and above. There is a slight amplification of jitter above 30 Hz owing to the time delays in pointing error signals and servo phase properties, as expected by the model calculation. When the servo performance was measured with the on-orbit operating configuration of SOT in the laboratory environment, the measured pointing error

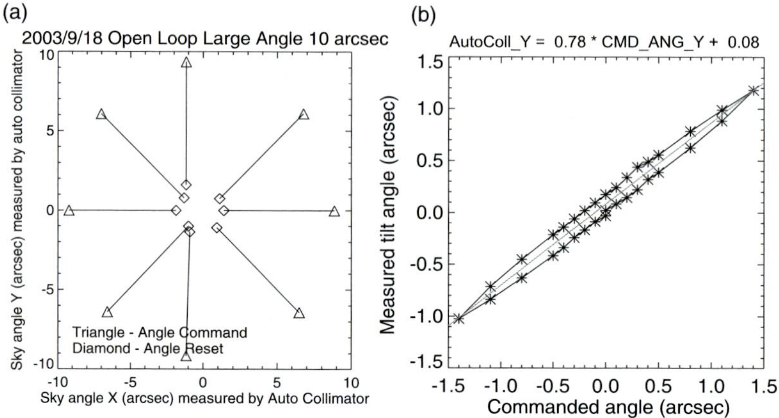

Figure 3 Nature of PZT hysteresis: (a) Home position uncertainty at tip-tilt angle resets from eight extreme positions near the edge of the stroke range. (b) Measured tilt angle as a function of angle commanded in ± 1.5 arcsec range.

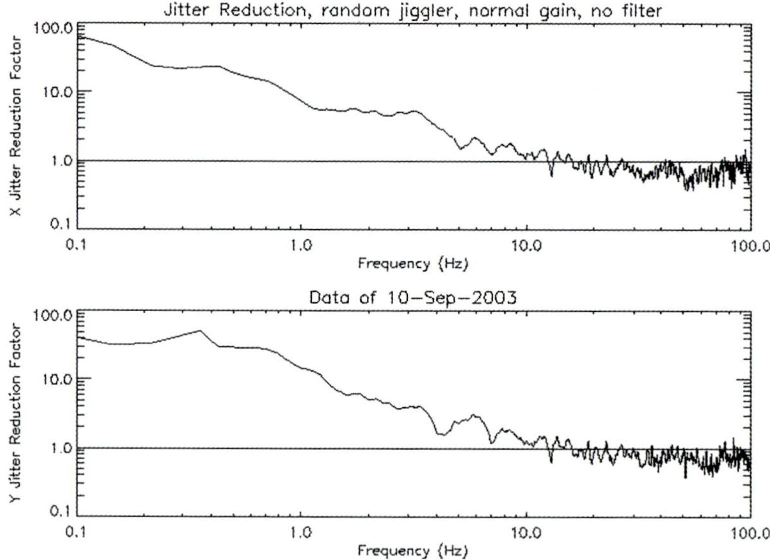

Figure 4 Jitter reduction for random image motion. The nano-positioner moved the target slide with a random noise input in *X* and *Y* over the $0-25$ Hz frequency band.

with the servo loop closed was 0.078 arcsec (3σ) in one direction and 0.038 arcsec (3σ) in the orthogonal direction. It should be noted again that this measured jitter error contains a significant contribution from the ambient jitter environment in the laboratory. From the measured power spectrum density plot, it can be derived that the error residuals in the $0-20$ Hz range are $0.003-0.006$ arcsec (3σ) when the servo is closed.

Before combining the correlation tracker and the tip-tilt mirror for the flight-model image stabilization system, other laboratory measurements were performed to evaluate how large errors can be induced by each portion of the system. The tip-tilt mirror mechanism and

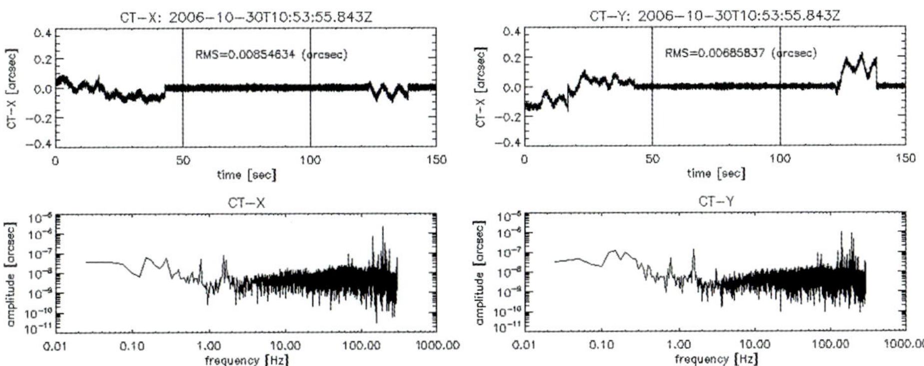

Figure 5 Time profiles and power spectral densities of the residual signal from the correlation tracker. The servo is closed during the time from 43 to 122 seconds, where the residual jitter is smaller than 0.026 arcsec (3σ). The power spectral densities are for $50-100$ seconds of the closed-loop control period.

its servo controller were evaluated by using a low-noise position-sensitive detector (PSD) with a laser point source on a floating optical bench, where ambient jitter was low with fewer resonances owing to more robust mechanical test supports. The measured jitter over the entire frequency range was $0.008-0.014$ arcsec (3σ), which indicates that the electrical noise induced in the tip-tilt mirror and its servo controller is the same or possibly lower than this level. The signals from the correlation tracker contain errors from photon noise and camera electrical noise and errors from the image correlation. The photon noise depends on the number of photons, and the error in the signals from the correlation tracker was designed to be smaller than 0.018 arcsec (3σ). Thus, the root-squared sum (RSS) of the pointing error for the combined system is calculated to be $0.020-0.023$ arcsec (3σ).

After *Hinode*'s launch, the on-orbit performance of the system has been evaluated, and excellent performance has been achieved, as expected from the pre-launch laboratory evaluation. Figure 5 shows time profiles and power spectral densities of the residual signal from the correlation tracker measured during the initial performance checking after the launch. The correlation tracker tracks a quiet region located at the solar disk center. The servo is closed in the time period 43 to 122 seconds. In this period the residual jitter is 0.026 arcsec (3σ) or less, which is much lower that the 0.09 arcsec (3σ) requirement. The power spectral densities are for $50-100$ seconds of the closed-loop control period. Two peaks are observed in the power spectra between $1-2$ Hz and just below 1 Hz. They are possibly excited by the small vibrations of the solar array paddles whose resonance frequency is 0.8 Hz. These two external solar panels (4300×1100 mm each) are attached at opposite sides of the spacecraft main structure that includes the telescopes.

We should note that other small peaks may be seen at 1.25 and 2.5 Hz when the CT tracks a magnetic-field-rich region. For polarization measurements, the polarization modulator located in the optical path is continuously rotated with a 1.6-second cycle. This rotating speed generates modulations of the circular polarization (Stokes V) signal with 1.25 Hz (0.8 seconds) and the linear polarization (Stokes Q and U) signal with 2.5 Hz (0.4 seconds). In the pre-launch laboratory evaluation, partially linear polarized light was fed to the CT with and without the rotation of the polarization modulator to evaluate how significant a pointing error might be generated by the polarization modulations. This measurement gave an increase of the pointing error at 2.5 Hz from 0.0027 arcsec (3σ) without the rotation to 0.0219 arcsec (3σ) with the rotation.

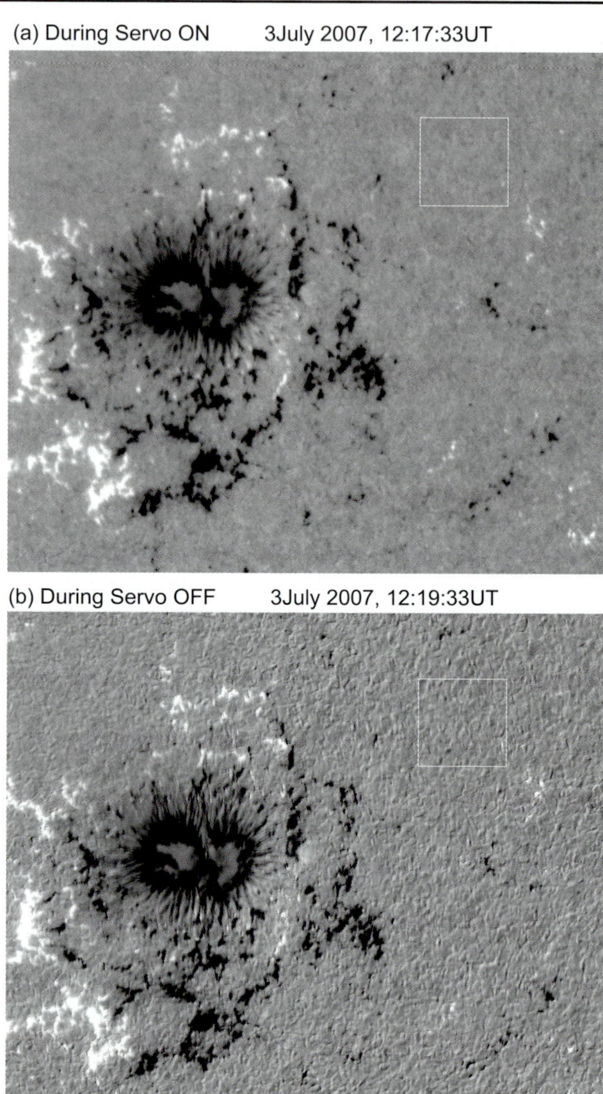

Figure 6 Longitudinal magnetograms acquired with servo on (a) and servo off (b). NFI shuttered Stokes *I* and *V* measurements (2×2 pixel summing, 0.16 arcsec) of active region NOAA 10961 were made using the Na I D line on 3 July 2007. The rectangular area given in the figures is used in the text to derive noise levels.

An example of scientific data from an SOT observation is shown in Figure 6 to demonstrate how crucial the image stabilization system is for achieving highly accurate polarimetric measurements. These are longitudinal magnetograms of the Na I D 589.6 nm line, which are derived by combining the filtergrams from multiple numbers of shuttered exposures. Figure 6a is the magnetogram acquired when the system servo control was on, whereas Figure 6b was acquired when the servo control was off. This comparison clearly shows the dramatic difference the servo makes. For a quiet area (Stokes *I* averaged intensity $=$ 8340 DN), the servo-on image shows 21 DN (σ) noise (S/N $=$ 400), whereas the servo-off image shows 50.5 DN (σ) noise (S/N $=$ 165). Thus, turning the servo off creates a new noise source with 46 DN (σ). Intensity gradients of granules easily produce false polarized signals when the image stabilization system does not remove the displacement among the combined images. It should be noted that the Spectro-Polarimeter, which measures the full Stokes spectra of magnetically sensitive Fe I 631.5/632.5 nm lines, requires polarimetric measurement accuracy of 0.1% or better. Thus, the Spectro-Polarimeter records two spectra simultaneously in orthogonal linear polarizations for further reduction of crosstalk between pointing errors and polarization.

5. Conclusions

An image stabilization system was developed for the *Hinode* Solar Optical Telescope. The performance of the flight system has been evaluated in the laboratory environment before the launch as well as on orbit after the successful launch, showing that the closed-loop system works as designed. The excellent performance provides extremely stable pointing for the SOT and makes a major contribution to precise polarimetric measurements with diffraction-limited images.

Acknowledgements   The image stabilization system was jointly developed by collaboration of Japanese and US instrument teams. The main body of the Japanese team is from the National Astronomical Observatory of Japan and Mitsubishi Electric Corporation and the US team is from the Lockheed Martin Advanced Technology Center Solar and Astrophysics Laboratory. The authors would like to express their thanks to all the engineers, technicians, scientists, and contributors involved in this development, especially Makoto Endo, Toshitaka Nakaoji, Norimasa Yoshida, Shiro Miki, Kazuhiro Otsuki, Tadashi Matsushita, Tetsuya Adachi, Noboru Kawaguchi, and Hideo Saito of Mitsubishi Electric Corporation, Kazumasa Kaneko and Shusaku Inoue of Mitsubishi Space Software Co., Ltd., and Hideaki Sawada, Shigekazu Muraki, and Hiroyuki Miyagawa of Systems Engineering Consultants Co., Ltd., for their deep contributions to the development of the tip-tilt mirror mechanism (CTM-TM) and its control electronics (CTM-E and CTM-TE), Tom Cruz, Gary Kushner, and Dnyanesh Mathur of Lockheed Martin Advanced Technology Center for the development of the correlation tracker in FPP, and Yasushi Sakamoto of the University of Tokyo and Masashi Otsubo, Kazuyoshi Kumagai, Motokazu Noguchi, Tomonori Tamura, Masakuni Miyashita, Yoshihiro Kato, and Masao Nakagiri of the National Astronomical Observatory of Japan for their ground test support, and Yukio Katsukawa for his leading efforts during the last stage of spacecraft-level final checking and on-orbit operations. Also, because no space-qualified PZTs are available in the world, a commercial product manufactured by Queensgate Instruments, Ltd., was qualified by ourselves for use in space with fruitful suggestions by Hitoshi Ariu, Paul Atherton, Chris Pietraszewski, Keith Gambles, and Angela Purnell of Queensgate Instruments, Ltd. Luc Thepaut, Eric Ruth, and Renaud Mercier Ythier of SAGEM/REOSC are also acknowledged for manufacturing the flight mirror for the tip-tilt mechanism. The authors also express their thanks to Takeo Kosugi of ISAS/JAXA and Larry Hill of NASA Marshall Space Flight Center for their encouragement and suggestions from the viewpoint of project management.

References

Ballesteros, E., Collados, M., Bonet, J.A., Lorenzo, F., Viera, T., Reyes, M., Rodríguez, Hidalgo, I.: 1996, *Astron. Astrophys.* **115**, 353.

Handy, B.N., Acton, L.W., Kankelborg, C.C., Wolfson, C.J., Akin, D.J., Bruner, M.E., *et al.*: 1999, *Solar Phys.* **187**, 229.

Ichimoto, K., Lites, B., Elmore, D., Suematsu, Y., Tsunete, S., Katsukawa, Y., *et al.*: 2007, *Solar Phys.*, submitted.

Kosugi, T., Matsuzaki, K., Sakao, T., Shimizu, T., Sone, Y., Tachikawa, S., *et al.*: 2007, *Solar Phys.* **243**, 3.

Molodij, G., Rayrole, J., Madec, P.Y., Colson, F.: 1996, *Astron. Astrophys. Suppl.* **118**, 179.

Shimizu, T.: 2005, In: Sakurai, T., Sekii, T. (eds.) *The Solar-B Mission and the Forefront of Solar Physics*, *Astron. Soc. Pacific Conf. Ser.* **325**, 3.

Shimizu, T., Nagata, S., Edwards, C., Tarbell, T., Kashiwagi, Y., Kodeki, K., *et al.*: 2004, In: Mather, J.C. (ed.) *Optical, Infrared, and Millimeter Space Telescopes*, *Proc. SPIE* **5487**, 1199.

Suematsu, Y., Tsuneta, S., Ichimoto, K., Shimizu, T., Otsubo, M., Katsukawa, Y., *et al.*: 2007, *Solar Phys.*, submitted.

Takahara, O., Yoshida, N., Minesugi, K., Hashimoto, T., Ninomiya, K., Ichimoto, K., *et al.*: 2004, In: *Proceedings of the 24th International Symposium on Space Technology and Science (CDROM)*, The Japan Society for Aeronautical and Space Science, paper 2004-C-16.
Tarbell, T.D., *et al.*: 2007, *Solar Phys.*, submitted.
Tsuneta, S., Suematsu, Y., Ichimoto, K., Shimizu, T., Otsubo, M., Nagata, S., *et al.*: 2007, *Solar Phys.*, submitted.
van der Lühe, O., Widener, A.L., Rimmele, T., Spence, G., Dunn, R.B., Wiborg, P.: 1989, *Astron. Astrophys.* **224**, 351.

Polarization Calibration of the Solar Optical Telescope onboard *Hinode*

K. Ichimoto · B. Lites · D. Elmore · Y. Suematsu · S. Tsuneta · Y. Katsukawa ·
T. Shimizu · R. Shine · T. Tarbell · A. Title · J. Kiyohara · K. Shinoda · G. Card ·
A. Lecinski · K. Streander · M. Nakagiri · M. Miyashita · M. Noguchi · C. Hoffmann ·
T. Cruz

Originally published in the journal Solar Physics, Volume 249, No 2.
DOI: 10.1007/s11207-008-9169-9 © Springer Science+Business Media B.V. 2008

Abstract The Solar Optical Telescope (SOT) onboard *Hinode* aims to obtain vector magnetic fields on the Sun through precise spectropolarimetry of solar spectral lines with a spatial resolution of $0.2 - 0.3$ arcsec. A photometric accuracy of 10^{-3} is achieved and, after the polarization calibration, any artificial polarization from crosstalk among Stokes parameters is required to be suppressed below the level of the statistical noise over the SOT's field of view. This goal was achieved by the highly optimized design of the SOT as a polarimeter, extensive analyses and testing of optical elements, and an end-to-end calibration test of the entire system. In this paper we review both the approach adopted to realize the high-precision polarimeter of the SOT and its final polarization characteristics.

Keywords Polarimeter · Stokes vector · Space telescope · Magnetic field · Optical telescope · Sun

K. Ichimoto (✉) · Y. Suematsu · S. Tsuneta · Y. Katsukawa · K. Shinoda · M. Nakagiri · M. Miyashita ·
M. Noguchi
National Astronomical Observatory of Japan, 2-21-1, Osawa, Mitaka, Toyko 181-8588, Japan
e-mail: ichimoto@solar.mtk.nao.ac.jp

B. Lites · D. Elmore · G. Card · A. Lecinski · K. Streander
High Altitude Observatory, National Center for Atmospheric Research, P.O. Box 3000 Boulder, CO
80307-3000, USA

R. Shine · T. Tarbell · A. Title · C. Hoffmann · T. Cruz
Lockheed Martin Advanced Technology Center, 3251 Hanover Street, Palo Alto, CA 94304, USA

T. Shimizu
Japan Aerospace Exploration Agency, Institute of Space and Astronautical Science, 3-1-1, Yoshinodai,
Sagamihara, Kanagawa 229-8510, Japan

J. Kiyohara
Kwasan Observatory, Kyoto University, Kita-Kazan Ohmine-cho, Yamashina-ku, Kyoto 607-8471,
Japan

1. Introduction

The science goals of the Solar Optical Telescope (SOT; Tsuneta *et al.*, 2008) onboard *Hinode* (formerly *Solar-B*; Shimizu, 2004; Ichimoto *et al.*, 2005; Kosugi *et al.*, 2007) require high-precision polarimetry of solar spectral lines with a spatial resolution of $0.2-0.3$ arcsec. *Hinode*/SOT thus provides the first quantitative and continuous measurements of full vector magnetic fields of the Sun that either resolve or isolate the solar fine-scale magnetic structures. The Focal Plane Package (FPP) of the SOT contains two sets of vector magnetographs (Tarbell *et al.*, 2008). The Spectropolarimeter (SP) performs the highest precision polarimetry with a photometric accuracy of $\approx 10^{-3}$ to provide full Stokes profiles of Fe I 630.25/630.15 nm lines with a spatial sampling of 0.16 arcsec. The Narrowband Filter Imager (NFI) of Filtergraph (FG), in contrast, produces two-dimensional images of Stokes parameters using a Lyot-type tunable filter (width ≈ 0.1 Å) in several photospheric and chromospheric lines with spatial sampling of $0.08''$/pixel and higher time cadence, but with lower wavelength resolution. The available spectral bands of the NFI contain Fe I 630.2/630.1/525.0/524.7 nm for photospheric magnetograms, Na I 589.6/Mg I 517.2 nm for chromospheric magnetogram/Dopplergrams, Fe I 557.6 nm for photospheric Dopplergrams, and H I 656.3 nm for chromospheric images/Dopplergrams. Both the SP and the NFI have a field of view (FOV) of 328×164 arcsec^2.

One of the most significant sources of error in high-spatial-resolution ground-based solar polarimetry is noise caused by atmospheric seeing. Since seeing produces rapid image motion, blurring, and distortion, if the polarization modulation is slower than 1000 Hz, seeing causes false "polarization" signals. Furthermore, attaining high spectropolarimetric precision (10^{-3} relative to the continuum intensity, I_c) at the telescope resolution demands integration times of at least several seconds. Even with adaptive optical correction, atmospheric seeing can significantly degrade image quality. As a result, an accuracy of 10^{-3} in Stokes vector measurements has rarely been achieved at a spatial resolution of less than 1 arcsec, and then never for an extended period of time. *Hinode*/SOT overcomes this difficulty by flying the telescope in space and stabilizing the residual pointing error with an image stabilization system (Shimizu *et al.*, 2004, 2008). The next major source of the error in polarization measurement is the instrumental polarization. Most large ground-based solar telescopes employ feed optics with oblique, time-varying reflective angles, which introduce considerable time variation in the instrumental polarization. In contrast, the SOT consists of symmetric optical system with constant, small-angle reflections. Since the entire observatory (satellite) points to the Sun, the instrumental polarization of *Hinode*/SOT is small and nearly constant. However, because on-orbit polarization calibration of the instruments and telescope is impractical, and because the SOT is exposed to significant thermal variation, a major design effort and comprehensive polarization tests of the system were required prior to launch.

In this paper, we review the methodology used for calibrating the SOT polarization and describe the final characteristics of the SOT polarimeter. The overview of the SOT as a polarimeter is described in Section 2, the goal of polarization calibration accuracy is defined quantitatively in Section 3, some component-level calibration tests are described in Section 4, and system calibration using the final SOT configuration is described in Section 5. Characterization and modeling of SOT polarization is discussed in Section 6 with additional information on data sampling schemes in Section 7. Section 8 summarizes the conclusions.

2. Overview of the SOT Polarimeter

Figure 1 presents a schematic diagram of the SOT optical system, emphasizing those components essential to polarimetry. The Optical Telescope Assembly (OTA; Ichimoto *et al.*, 2004; Suematsu *et al.*, 2008) is a 50-cm-aperture Gregorian telescope containing the primary mirror, secondary mirror, collimator lens unit (CLU), polarization modulator unit (PMU), tip-tilt folding mirror (CTM-TM), and astigmatism corrector lens (ACL) as elements that may act on the polarization states of the incident light. The primary and secondary mirrors have ellipsoidal figures and protected silver coatings. The CLU is an achromatic lens unit that consists of six elements of different types of optical glasses. The CLU provides a collimated beam to the FPP and creates a 30-mm ϕ pupil image between the CTM-TM and the ACL. The ACL is a nearly plane parallel silica plate with a thickness of 10 mm. This plate was installed after assembly and testing of the Gregorian telescope to eliminate the as-built small astigmatism of the telescope (see Suematsu *et al.*, 2008). To prevent a ghost from a reflection of the collimated beam, the plate was tilted by 1.5 degrees with respect to the optical axis. The PMU is a bicrystalline athermal waveplate that rotates at a constant rate of 1/1.6 Hz. Retardation is optimized for measurements of both circular and linear polarization at 630.2 and 517.2 nm (Guimond and Elmore, 2004). The beam is split between the SP and the FG paths by a nonpolarizing beam splitter. The polarizer in front of the tunable filter and a polarizing beam splitter in front of the SP CCD provide the polarization analysis for FG/NFI and SP, respectively.

Both SP-CCD and FG-CCD take multiple images synchronously with the PMU. The SP takes 16 exposures for each PMU revolution in both orthogonally polarized beams. The FG has a variety of sampling schemes. In the "shuttered mode" of the NFI, the mechanical shutter is used to control the exposure of a large area of the CCD. The mechanical shutter

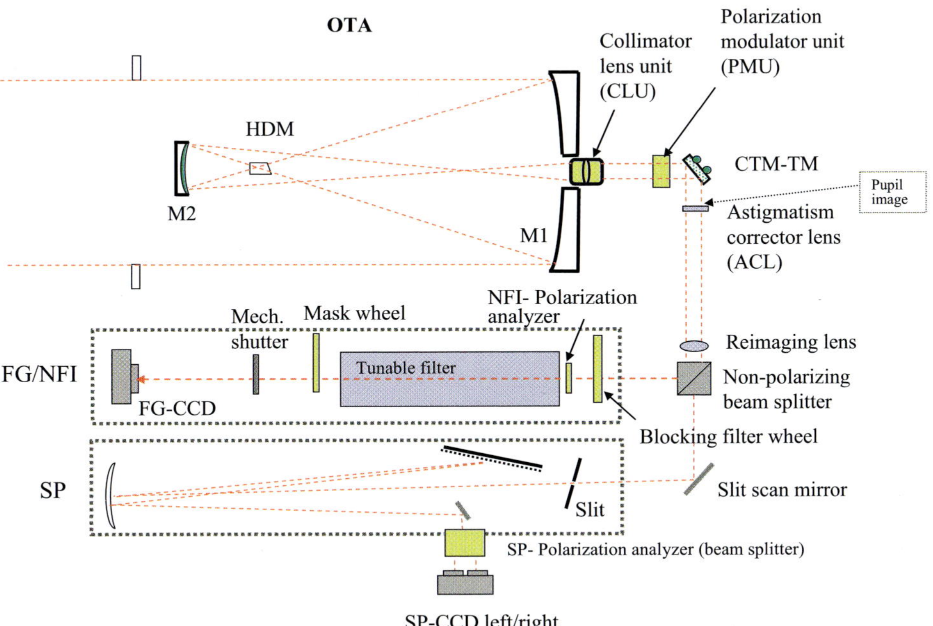

Figure 1 Schematic of the SOT polarimeter.

is placed at a pupil image to avoid PMU phase variation across the field of view during exposures. In "shutterless mode," continuous readout is performed in the central area of the CCD. The mask wheel located at an intermediate focus has a selectable aperture that masks off different amounts of the outer portions of the CCD. Appropriate demodulation (successive addition and subtraction of images) is applied onboard for each sampling sequence to reduce data telemetry volume data and to improve signal to noise (S/N). The SOT data products are then I, Q, U, and V images, which require further correction by a calibration matrix to obtain the Stokes vector of the light incident to the telescope. Typical sampling schemes of the NFI are as follows:

In shuttered mode:
    8 exposures at 22.5° waveplate rotation intervals for $IQUV$.
    4 exposures for $IQUV$.
    2 exposures for IV.
Other exposure schemes are also possible with variable exposure times.
In shutterless mode:
    16 frames/rev. for $IQUV$ (same as SP).
    4 frames/half rev. for IV.
    2 frames/half rev. for IV.
Other exposure schemes are also possible with flexible numbers of accumulations.

The key features that optimize the SOT as a polarimeter are summarized as follows:

1. Axisymmetric configuration up to the polarization modulator: As shown in the next section, the optical system up to the polarization modulator is most critical to the accurate measurement of polarization. The axisymmetric configuration of this system is a great advantage for minimizing the crosstalk among the Stokes $IQUV$.
2. Simple rotating waveplate for the polarization modulator: Since a rotating waveplate causes Stokes Q, U, and V to be modulated at different frequencies and phases, the crosstalk among them is not sensitive to the absolute retardation of the waveplate.
3. Rotating waveplate located near the pupil image: Any possible defect or nonuniformity of the waveplate will not produce spurious intensity modulation at the detector.
4. Simultaneous measurements of both orthogonally polarized beams at the SP-CCD: Any residual guiding error of the spacecraft will produce an intensity modulation. This term (I to QUV crosstalk) may be greatly reduced by combining observables taken simultaneously in the two orthogonal polarizations.

3. Requirement on Accuracy of the Polarization Calibration

The "polarimeter response matrix" $\mathbf{X}$ (Elmore, 1990) is defined as $\mathbf{S}' = \mathbf{X}\mathbf{S}$, where $\mathbf{S}$ is the incident Stokes vector to the telescope and $\mathbf{S}'$ is the data product of the SOT (demodulated intensity; see Figure 2). Our goal of the polarization calibration of the SOT is to determine the $\mathbf{X}$ matrix of both the SP and the FG/NFI with a required accuracy as described in the following. For the crosstalk among different elements of the Stokes vector, we require that a fictitious signal produced by the incorrect evaluation of $\mathbf{X}$ should be smaller than the statistical noise (photon noise). Denoting the polarimeter response matrix used in data reduction as $\mathbf{X}_r$, we can write the error in the reduced Stokes vector as

$$\Delta\mathbf{S} = \mathbf{S}'' - \mathbf{S} = \mathbf{X}_r^{-1}\mathbf{S}' - \mathbf{S} = \left(\mathbf{X}_r^{-1}\mathbf{X} - \mathbf{E}\right)\mathbf{S},$$

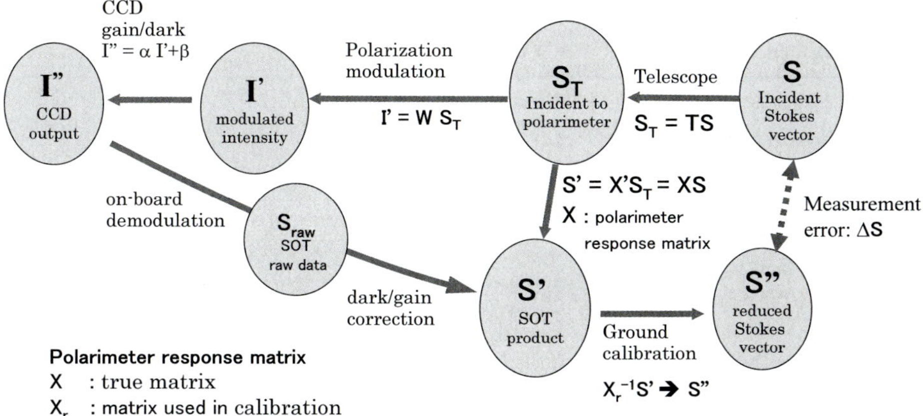

Figure 2 Definition of polarization response matrix and error in polarization measurements.

where $\mathbf{S}''$ is the reduced Stokes vector, $\mathbf{X}$ is the real, but unknown, polarimeter response matrix, and $\mathbf{E}$ is the identity matrix. The error in the reduced Stokes vector from the statistical noise, ε, is given by

$$\delta\mathbf{S} = \mathbf{X}_r^{-1}\delta\mathbf{S}' = \mathbf{X}_r^{-1}\boldsymbol{\varepsilon},$$

where $\boldsymbol{\varepsilon}$ is a four-element column vector with all elements having a value of ε.

The requirement of $\Delta\mathbf{S} < \delta\mathbf{S}$ reduces to

$$\Delta\mathbf{X}\mathbf{S} < \boldsymbol{\varepsilon},$$

where $\Delta\mathbf{X} \equiv (\mathbf{X} - \mathbf{X}_r)$ is the required accuracy for $\mathbf{X}$. This inequality is interpreted as follows: Let p_l and p_c denote the maximum linear and circular polarizations, respectively, in realistic spectral lines from the Sun. When one applies $\Delta\mathbf{X}$ to the Stokes vectors representing the anticipated maximum polarization (*i.e.*, $\mathbf{S} = (1, p_l, 0, 0)^T$, $(1, 0, p_l, 0)^T$, or $(1, 0, 0, p_c)^T$ incident to the telescope), the resulting error of each Stokes parameter should be smaller than ε. In particular, off-diagonal elements of $\Delta\mathbf{X}$ produce a false signal of a Stokes component even when the component is intrinsically zero via the crosstalk from other Stokes components. Such errors must be suppressed below the detection limit of the system (*i.e.*, ε). However, the "scale error," which is introduced by the error of diagonal elements or the first row of $\mathbf{X}$, does not produce a false signal of a Stokes component if that component is intrinsically zero, but changes its value with a certain factor. Since the scale error exists also in derivations of the magnetic fields from the Stokes profiles of spectral lines because of uncertainties in solar atmospheric models, we relax the requirement on the scale errors and set the limit by an "uncertainty factor" a rather than ε. After normalizing $\mathbf{S}$ by the intensity ($I = 1$ and $x_{11} = 1$, we can write the inequality for $\Delta\mathbf{X}$ as

$$|\Delta\mathbf{X}| < \begin{pmatrix} - & a/p_l & a/p_l & a/p_v \\ \varepsilon & a & \varepsilon/p_l & \varepsilon/p_v \\ \varepsilon & \varepsilon/p_l & a & \varepsilon/p_v \\ \varepsilon & \varepsilon/p_l & \varepsilon/p_l & a \end{pmatrix}.$$

Table 1 Classification of polarimeter components.

	Definition	SOT components	
M_T	Elements before polarization modulator	Gregorian telescope, CLU	
M_P	Polarization modulator	Rotating waveplate (PMU)	
M_B	Elements between polarization modulator and analyzer	Tip-tilt mirror, reimaging lens, beam splitter	
		(SP)	(NFI)
		Scanning mirror	Folding mirrors
		Blocking filter	
		Slit	
		Field lens	
		spectrograph	
M_A	Polarization analyzer	Polarizing beam splitter	Polarizer
M_F	Elements following the polarization analyzer	CCD	Filters
			Relay lenses
			Folding mirror
			CCD

In case of *Hinode*/SOT, we adopt the following values:

$$\varepsilon = 0.001,$$

$$a = 0.05,$$

$$P_l = 0.15 \quad (\text{max of } Q, U),$$

$$P_c = 0.2 \quad (\text{max of } V),$$

and hence the tolerance of $\mathbf{X}$ becomes

$$|\Delta\mathbf{X}| < \begin{pmatrix} - & 0.333 & 0.333 & 0.250 \\ 0.001 & 0.050 & 0.007 & 0.005 \\ 0.001 & 0.007 & 0.050 & 0.005 \\ 0.001 & 0.007 & 0.007 & 0.050 \end{pmatrix}.$$

This relation gives the basis of our goal of the SOT polarization calibration.

The optical components in a polarimeter can be classified into five groups based on their location in the optical system with respect to the polarization modulator and the analyzer. Table 1 shows the category of the polarization elements in the SOT. The tolerances of polarization properties of components in each group may be evaluated using the tolerance matrix $\Delta\mathbf{X}$ specified here; that is, one may calculate the amount of error in diattenuation, retardation, or depolarization of each of the elements that causes the error of one of the elements of $\mathbf{X}$ to exceed the tolerance $\Delta\mathbf{X}$. Table 2 shows these calculated tolerances of the polarization properties for each group.

It is obvious from Table 2 that the first elements in the optical train have the tightest tolerance limits and components in the OTA need to be carefully characterized for their

Table 2 Tolerance of polarization properties for each component.

Location	Diattenuation	Retardation (deg)	Orientation (deg)	Depolarization
M_T	$0.0010\ (I \to Q, U)$	$0.286\ (V \to U)$	Does not matter	$0.050\ (dQ, U, V)$
M_P	$0.0053\ (U \to V)$	$3.687\ (dV)$	$0.095\ (Q\text{–}U)$	$0.050\ (dQ, U, V)$
M_B	$0.0073^*\ (Q\text{–}U)$	$0.419^*\ (Q \to V)$	$0.100\ (U\text{–}Q)$	$0.050\ (dQ, U, V)$
M_A	Does not matter	Does not matter	$0.233\ (Q\text{–}U)$	Does not matter
M_F	Does not matter	Does not matter	$2.100\ (Q\text{–}U)$	Does not matter

Note. The kind of crosstalk that limits the tolerance of each error is shown in parentheses.

*The axis of error is assumed to be $45°$ to the axis of the polarization analyzer. Off-axis rays from the edge of the FOV entering on the CTM-TM or the BS correspond to an axis rotation of $\approx 0.7°$.

polarization properties. In the measurements of **X** matrices of the SOT, the uniformity of the **X** matrices over the field of view and their temperature stability are also important aspects to be characterized. In the next section, we focus on polarization calibration of the critical components of the OTA (*i.e.*, optical coatings and the CLU).

4. Component Calibration

In the development of the SOT, the highest priorities for design of optical components and selection of materials were their durability in the space environment and ability to achieve high wavefront quality. Where choices were possible (*e.g.*, optical coatings, waveplate design, *etc.*), polarization properties were also a major factor for the selection. The expected polarization characteristics of each component in a realistic space thermal environment was studied based on theoretical properties in the design phase. After the fabrication of each component, polarization properties were measured and characterized using the Component Polarization Analyzer (CPA) developed by HAO. The CPA consists of a "polarization generator" and a "polarization analyzer" with a sample to be measured in between them. The former creates known polarized lights (a set of Stokes vectors) and the later measures the Stokes vectors of the light after passage through the sample. The spatial distributions of 16 elements of the Mueller matrix of the sample can be obtained as two-dimensional maps with an accuracy that meets our requirement. In this section, examples of component polarization calibration are described for representative cases that are the most critical for accurate polarimetry with the SOT (*i.e.*, optical coatings in the OTA and the CLU).

4.1. Optical Coatings in the OTA

Coatings on mirrors or lenses can directly affect the polarization state of the beam, and those in the OTA are especially critical for the final performance of the polarimeter. The primary and secondary mirrors are 1170- and 263-mm focal length ellipsoids with a protected silver coating provided by SAGEM/REOSC. The CLU is a six-element achromatic lens schematically drawn in Figure 3. All elements have an antireflection coating on their surfaces provided by Canon, except for the first surface, which has a bandpass coating to reject most IR and ultraviolet wavelengths. The coating was fabricated by the Ion Beam Spattering system of NAOJ. There are in total 14 optical surfaces before the light reaches the PMU. Figure 4 shows the maximum incidence angles of the rays from the center and

Figure 3 Configuration of CLU with coatings on each surface.

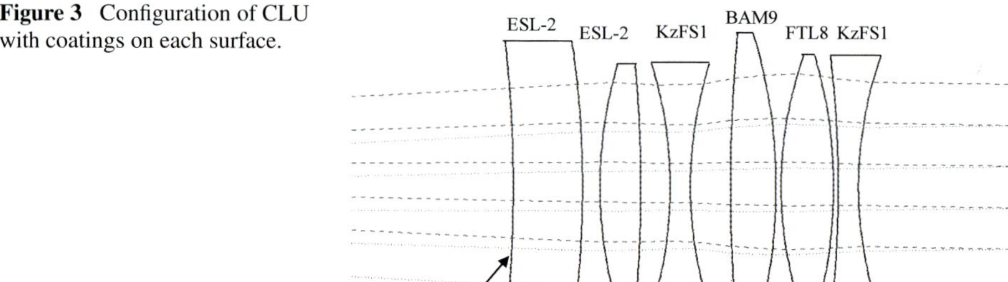

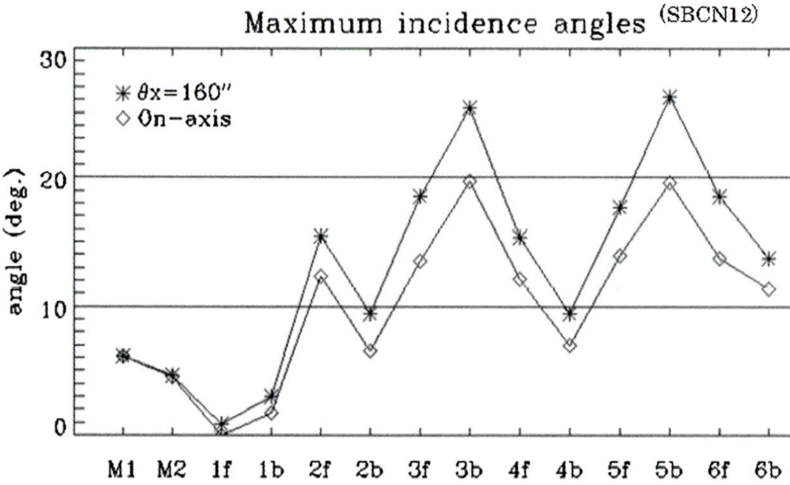

Figure 4 Maximum incidence angle of rays at each surface of M1, M2, and CLU.

edge of the field of view at each surface. Since the CLU is optically rather fast, these angles are up to 26° for particular rays at the edge of the FOV.

Figure 5 shows the theoretical polarization properties of the protected silver coating on M1 and M2 (left) and the antireflection coating of one of the CLU elements (right) as a function of incident angle. Shown from top to bottom are the transmission of P and S polarizations, diattenuation, and retardation. These properties were evaluated by using the actual measurements of coating witness samples by the CPA. Examples of the results are shown with error bars in the right panel of Figure 5. It should be remarked that the most significant outcome from the measurements is the confirmation that there is neither unexpected retardation nor diattenuation at normal incidence. Actually, we found that some coatings, which were deposited under anisotropic conditions, can have such behavior, and one coating that had an "intrinsic" polarization was rejected in the development of the SOT.

To evaluate the net polarization of the telescope from the optical coatings, a polarizing ray tracing was performed in which the propagation of the Stokes vectors (or Jones vectors) of individual rays were calculated based on the polarization properties of the coatings. The rays were combined after passing through the optics. Because of the axisymmetric configuration

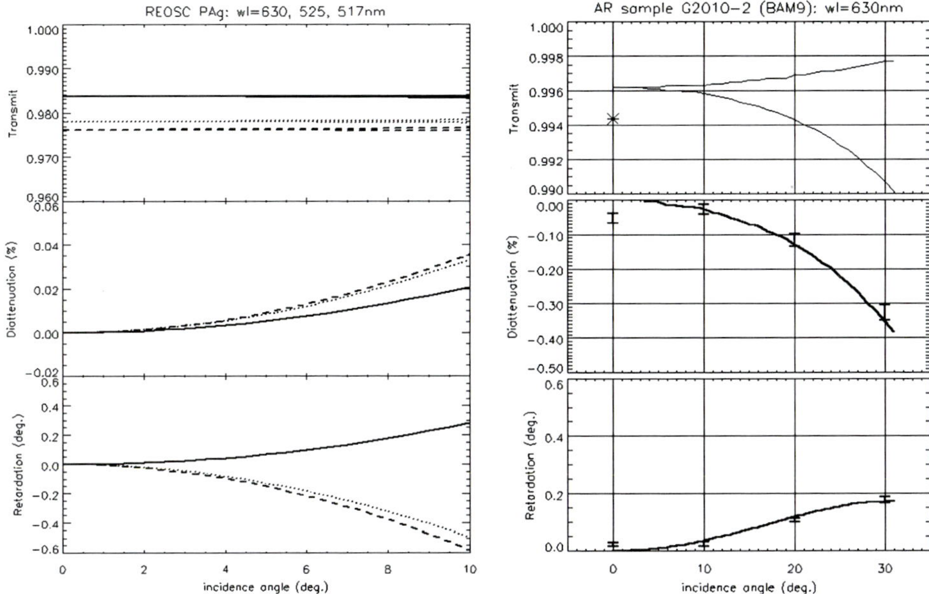

Figure 5　Theoretical polarization properties of the protected silver coating on M1 and M2 (left) and the antireflection coating of one of the CLU elements (right) as a function of incident angle: transmission of P and S polarizations (top), diattenuation (middle), and retardation (bottom). Shown are curves for a protected silver coating at 517 nm (dashed), 525 nm (dotted), and 630 nm (solid). In the right panel, measurement results are also shown with error bars.

of the telescope, the net polarization is theoretically zero at the center of the field of view, and the Gregorian telescope ($M1 + M2$) has diattenuation and retardation of the order of 10^{-8} and 10^{-4} degrees at the edge of FOV, respectively (both of which are much smaller than our tolerance). Figure 6 shows an example of calculations for the CLU at the edge of the FOV. Short lines correspond to the rays incident at different points on the pupil. The length and direction of each line show the amount and direction of the diattenuation resulting from the 12 surfaces of the CLU. After combining these rays, one obtains the net diattenuation and retardation of the CLU coatings: about 3×10^{-4} and $0.08°$, both of which, again, are smaller than our requirements given in Table 2.

4.2. CLU Optothermal Properties

The six elements of the CLU are tightly mounted in a titanium housing to maintain their precise positions and survive the launch load. (For details of the CLU design, see Suematsu *et al.* (2008).) In contrast to reflective surfaces, at the CLU, light passes through the optical media, which may have internal stress. Thus the CLU polarization may be quite sensitive to the properties of the optical glasses and also temperature since the mechanical stress on the glasses induced from the titanium housing may cause additional retardation. This led us to perform extensive analyses and testing to characterize the optothermal polarization properties of the CLU.

Prior to fabrication of the CLU, we measured the retardation of the glass blanks for all lens elements caused by the residual internal stress and also the stress-optical coefficients of the optical glasses using a HeNe laser at 632.8 nm. Table 3 summarizes the results of

Figure 6 Result of polarizing ray tracing for the CLU for the rays at the edge of the SOT field of view (see text for details).

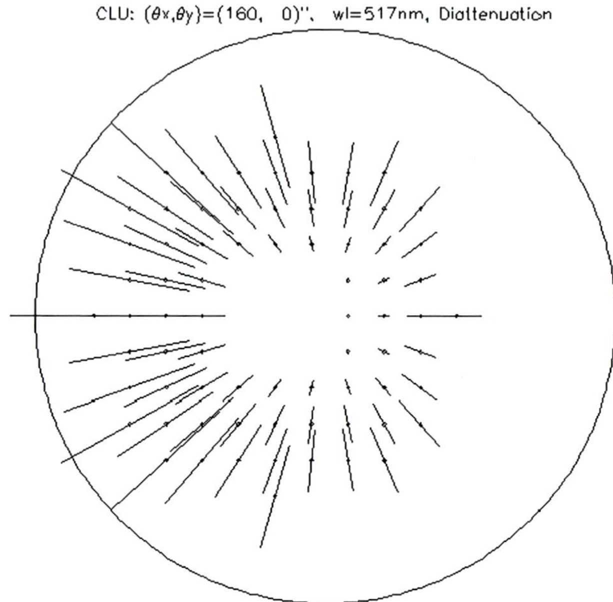

Table 3 Optomechanical properties of the CLU glasses (measurements by NAOJ).

	Retardation due to the internal stress (degree cm^{-1} @630 nm)	Stress-optical coeff. ($\times 10^{-6}$ mm^2 N^{-1})	Supplier
BAM9	0.033 – 0.035	1.89	Ohara
KzFS1	0.441 – 0.632	3.52	Schott
FTL8	0.087 – 0.123	3.17	Ohara
ESL2 (silica)	0.110 – 0.148	3.37	Tosoh

the measurements. We selected the glass blank that has a minimum retardation for the flight CLU, and, for the KzFS1, for which all blanks have non-negligible retardation, we aligned the axes of retardation of two KzFS1 elements in the CLU orthogonally to cancel the retardations. By performing extensive thermomechanical analyses, we calculated the internal stress distributions of each CLU element under realistic thermal conditions with the absorption of incident solar light, and, using the measured stress-optical coefficients, we evaluated the retardation of the CLU. The promising result indicated a small polarization effect of the CLU, but the simulation was based on an ideal axisymmetric mechanical model of the CLU.

Tests of polarization properties with the real flight CLU were performed by using the CPA. The CLU was mounted in a thermal shroud and put in a beam that simulates the flight optical configuration, and the Mueller matrix was obtained as a function of position in the FOV at $\lambda = 630.2$ nm under various temperature conditions before and after mechanical load and thermal cycling tests. Figure 7 shows examples of two-dimensional images of the Mueller matrix elements derived from the CPA measurement, where the temper-

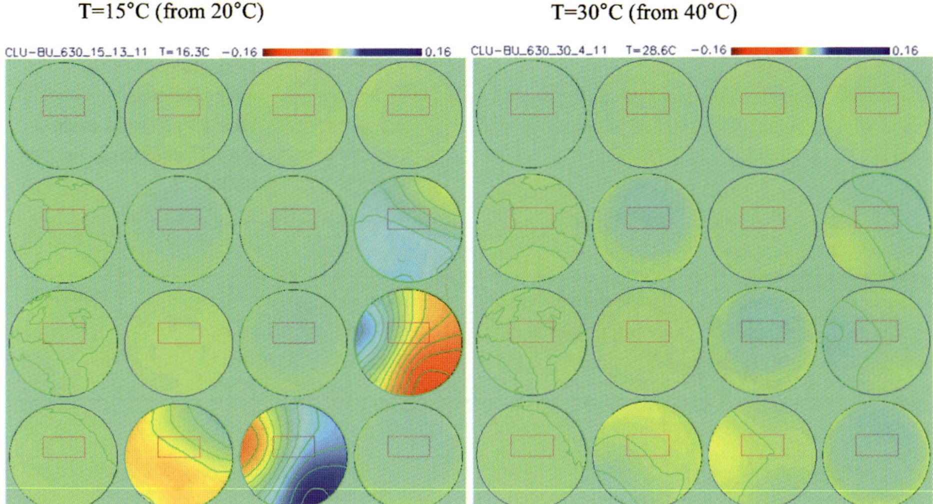

Figure 7 Mueller matrix image of the CLU at $T = 15°C$ (left) and $T = 30°C$ (right) as inferred from CPA measurements. Rectangles in each matrix element show the SOT field of view. The contour intervals indicate the tolerance of each Mueller matrix element.

ature of the CLU was 15°C (left) and 30°C (right). Each image spans different incident ray angles and the rectangle in each matrix element shows the SOT field of view. Contour intervals indicate the tolerance of each Mueller matrix element as defined in Section 3. We observe a clear indication of large, inhomogeneous polarization effects when the CLU temperature is low, especially in two elements in the last row ($[2, 4]$ and $[3, 4]$) and two elements in the last column ($[4, 2]$ and $[4, 3]$), which correspond to linear retardation. At $T = 15°C$, the retardation and its variation across the SOT field of view are significantly larger than our tolerance, as indicated by the contours. Such linear retardation may be created by lateral stress induced by the housing, since neither lenses nor housing are ideal owing to machining errors. It is also revealed that the behavior of the linear retardation has significant hysteresis. Figure 8 shows the history of the $[2, 4]$ element of the CLU Mueller matrix averaged over the SOT field of view versus temperature through various environmental tests.

After extensive measurements, we reached the following conclusions regarding the polarization of the CLU: The only significant polarization effect of the CLU is a linear retardation (*i.e.*, there are no linear and circular diattenuations, no circular retardation, and no depolarization). The CLU linear retardation may be regarded as uniform over the SOT field of view and constant against T at temperatures higher than 25°C. The CLU retardation can have a small, unpredictable shift following launch vibration and the initial cold cycle in orbit. The signature of circular to linear crosstalk will be checked after launch by using a simple sunspot, and that part of the calibration matrix will be updated if necessary. The lower limit of the operational temperature range of the CLU was thus set as 25°C, which will be maintained by the OTA operational heaters in orbit.

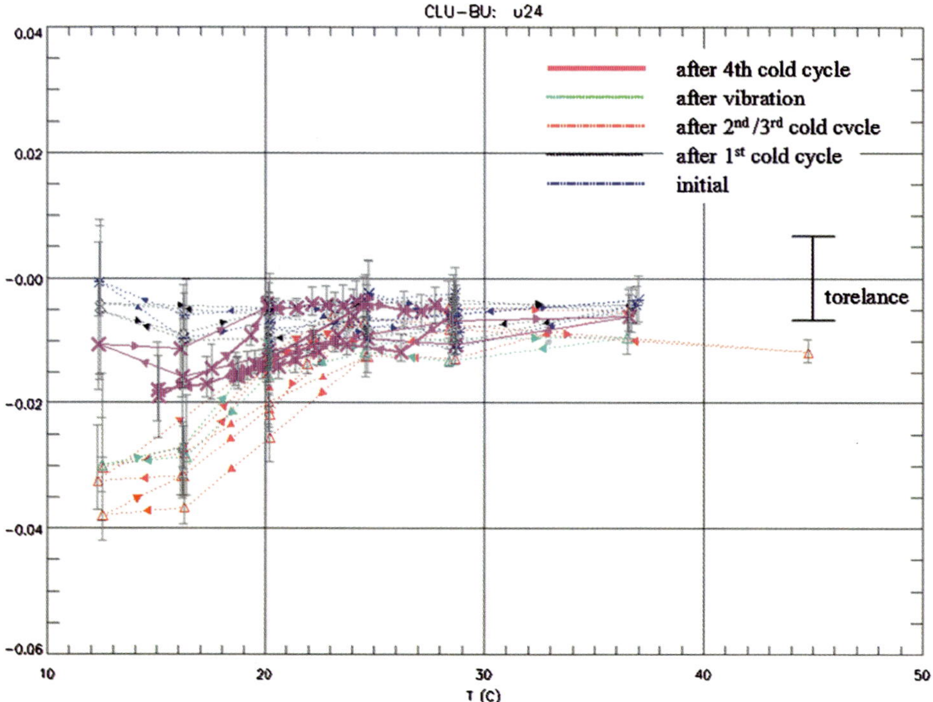

Figure 8 Hysteresis of $(2, 4)$ element (= linear retardation) of the CLU Mueller matrix against temperature. The averages over the SOT FOV are plotted with error bars showing the variation across the SOT FOV.

5. System Calibration

5.1. Definition of Coordinate System

Figure 9 shows the definition of the polarization coordinate system of the SOT with respect to the spacecraft coordinates. This definition follows the standard convention used in the data analysis of the ASP (Advanced Stokes Polarimeter; Skumanich and Lites, 1997); that is, right circular polarization is positive when the electric vector rotates clockwise looking at the source, and positive V on the blue side of spectral lines gives positive magnetic flux. Note that this definition is applied to the Stokes vectors obtained after application of the **X** matrix; therefore, the raw Stokes products of the SOT (which are also called $IQUV$) are not consistent with this definition.

5.2. Test Setup and Measurements

Measurement of the polarimeter response matrix of the SOT was performed by using natural sunlight fed by a heliostat at the NAOJ clean room (see Figure 10). The entire SOT (with the OTA and the FPP attached to the spacecraft Optical Bench Unit) was located under the heliostat. Well-calibrated sheet polarizers (linear, left, and right circular) were placed at the entrance of the telescope at $0°$, $45°$, $90°$, and $135°$. The polarizers create the incident Stokes vectors. At each position of the sheet polarizer, data were taken by both the SP and the NFI in typical observing sequences, with multiple sets of polarization products corresponding to

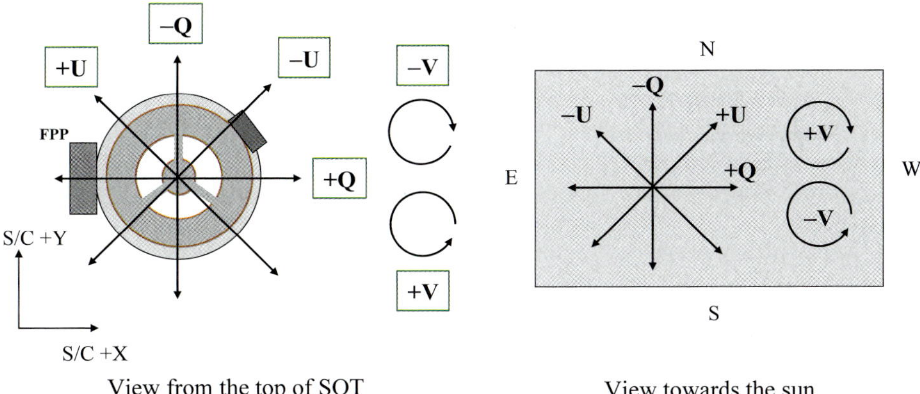

Figure 9 The definition of the SOT polarization coordinates.

Figure 10 Configuration of the SOT polarization calibration testing system.

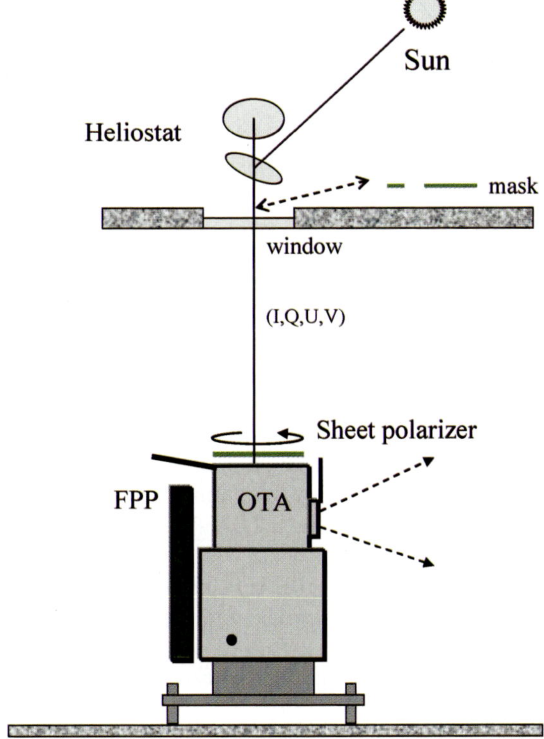

12 different incident Stokes vectors obtained. The data were taken for the entire slit scan range of the SP and at all available wavelengths using representative exposure schemes for the FG/NFI. During testing, the room temperature was controlled at 20°C, while the OTA operational heater was applied to maintain the CLU temperature at $T > 25°C$.

The sheet polarizers used for the test were HN38 (linear), HNCP37R (right-hand circular), and HNCP37L (left-hand circular), all provided by 3M Corporation. Their Mueller

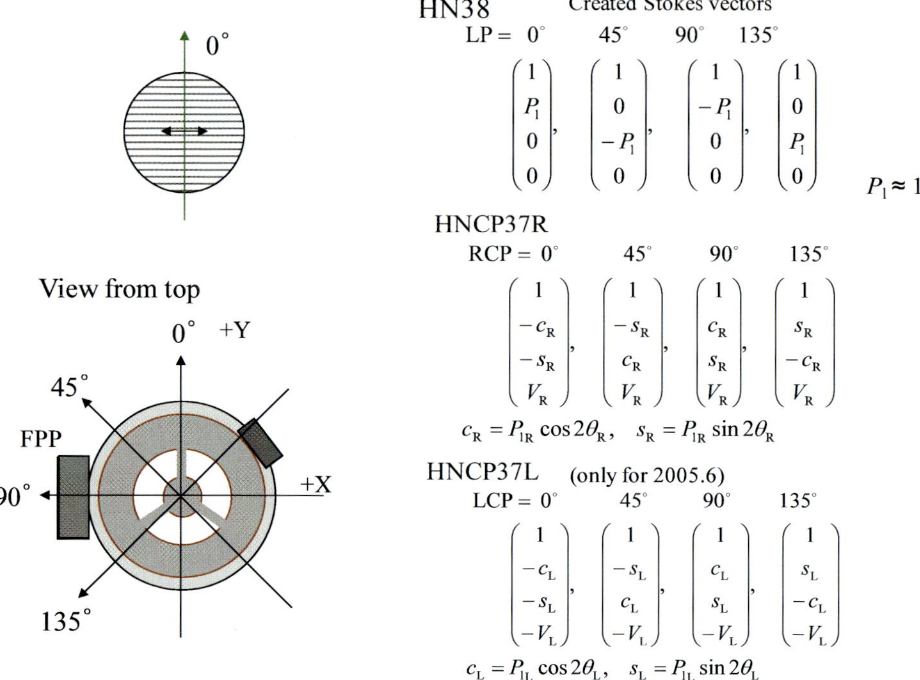

Figure 11 Configuration of sheet polarizer and incident Stokes vector.

matrices at 630 nm were obtained prior to testing by using the NAOJ Mueller matrix measurement system (Ichimoto *et al.*, 2006). The configurations of the sheet polarizer and corresponding incident Stokes vectors are shown in Figure 11. V_R and P_{1R} are circular and linear diattenuations of HNCP37R , and V_L and P_{1L} are circular and linear diattenuations of HNCP37L, respectively. We obtained values of $V_R = 0.9811$, $P_{1R} = 0.1496$, $V_L = 0.9905$, and $P_{1L} = 0.0637$ from the Mueller matrix measurements. The angles θ_R and θ_L are the orientations of the linear diattenuation of the HNCP37R and HNCP37L, respectively, and are regarded as unknown parameters in the following data reduction. It may be shown that the polarization produced by the heliostat ($< 5\%$) is negligible when these sheet polarizers are placed in front of the OTA.

Table 4 summarizes the data set taken during the two test periods for the SOT polarization calibration. The data sets marked with circles were used for the following analysis.

5.3. Derivation of the Polarimeter Response Matrix

The relation between the data products of the SP ($\mathbf{S}'$) and incident Stokes vectors ($\mathbf{S}$) may be written as follows:

$$\mathbf{S}_k'^{\pm} = \mathbf{X}^{\pm}\mathbf{T}\mathbf{S}_k \equiv \alpha^{\pm} I_k \mathbf{x}^{\pm}\mathbf{s}_k,$$

$$\begin{pmatrix} I_k' \\ Q_k' \\ U_k' \\ V_k' \end{pmatrix}^{\pm} = \alpha^{\pm} I_k \begin{bmatrix} 1 & x_{10} & x_{20} & x_{30} \\ x_{01} & x_{11} & x_{21} & x_{31} \\ x_{02} & x_{12} & x_{22} & x_{32} \\ x_{03} & x_{13} & x_{23} & x_{33} \end{bmatrix}^{\pm} \begin{pmatrix} 1 \\ q_k \\ u_k \\ v_k \end{pmatrix},$$

Table 4 List of data sets for the SOT polarization calibration.

λ	5172	5250	5896	6302	6563	Date	
FG shuttered $IQUV$							
(exp $=$ 90 ms)							
2048 $\times$ 1024 (2 $\times$ 2sum, OBS_ID $=$ 3)	–	◯	◯	◯	◯	2004.8.19/20	
512 $\times$ 1024 (2 $\times$ 2sum, OBS_ID $=$ 3)	(◯)	(◯)	(◯)	(◯)	(◯)	2005.6.13	Quality is poor (not used)
FG shutterless $IQUV$							
(exp $=$ 100 ms)							
64 $\times$ 2048 (1 $\times$ 1sum, OBS_ID $=$ 33)	–	◯	◯	◯	◯	2004.8.19/20	Mask $=$ 82
80 $\times$ 1024 (2 $\times$ 2sum)	–	◯	◯	◯	◯	2004.8.19/20	Mask $=$ 112
72 $\times$ 1024 (2 $\times$ 2sum)	◯	◯	◯	◯	◯	2005.6.13/14	Mask $=$ 112
SP							
224 $\times$ 1024 (1 $\times$ 1sum)				△		2004.8.19/20	SP opt. was modified later
				◯		2005.6.13/14	

where $\mathbf{S}'_k$ is the SP product, $\mathbf{s}_k$ is the normalized incident Stokes vector with $I = 1$, $\pm$ stands for the left and right CCD areas (measuring orthogonal polarizations), and k stands for the configuration of the sheet polarizer. Each element of $\mathbf{X}$ is determined by a least-squares fitting procedure by using the normalized equation by each I'_k to eliminate the variability of the sky transmission,

$$
\begin{pmatrix} Q'_k/I'_k \\ U'_k/I'_k \\ V'_k/I'_k \end{pmatrix} \equiv \begin{pmatrix} q'_k \\ u'_k \\ v'_k \end{pmatrix} = \frac{\begin{bmatrix} x_{01} & x_{11} & x_{21} & x_{31} \\ x_{02} & x_{12} & x_{22} & x_{32} \\ x_{03} & x_{13} & x_{23} & x_{33} \end{bmatrix}}{1 + x_{10}q_k + x_{20}u_k + x_{30}v_k} \begin{pmatrix} 1 \\ q_k \\ u_k \\ v_k \end{pmatrix}.
$$

The number of equations is thus $3 \times 12 \times 2 = 72$, while the number of unknowns is 15×2 (with $x_{00} = 1$). Fitting is carried out for each pixel of the CCD, but θ_R and θ_L (the offset angles of RCP and LCP) are determined prior to the fitting from the average over the CCD.

A similar approach is applied to obtain the $\mathbf{X}$-matrix elements of the FG/NFI, but only for one CCD, and the degree of circular/linear polarization of the circular polarizers is also regarded as unknown since we do not have Mueller matrix measurements of them at wavelengths other than 630.2 nm. The values of θ_R, θ_L, P_{lR}, and P_{lL} (the offset angles and linear polarizations of RCP and LCP) are determined from the average over the CCD and then fixed in fitting for each pixel by assuming $P^2_{cR} + P^2_{lR} = 1$ and $P^2_{cL} + P^2_{lL} = 1$. Thus, the number of unknowns is 15 and the number of equations is $3 \times 12 = 36$.

Thus the $\mathbf{X}$-matrix elements were determined as a function of the position in the FOV. We obtained two-dimensional maps of polarimeter response matrices for the SP at multiple scan positions covering its entire range and for the NFI in all available wavelengths for representative exposure schemes.

5.4. SP Polarimeter Response Matrix

It should be noted that the analyses of the SOT/SP polarization calibration measurements were carried out completely independently by two separate methods (at HAO and NAOJ). These methods produced the same results within the measurement error. The HAO calibration scheme has a heritage from calibration of the Advanced Stokes Polarimeter (Skumanich and Lites, 1997). It has been used to calibrate other ground-based polarimeters. Here we present results derived from the NAOJ scheme, but the calibration software for flight data currently utilizes data resulting from the HAO calibration.

Figure 12 shows an example of observed products of the SP along with fitting results for both left and right CCD areas. This result is for a particular pixel at the center of the CCD and for the center of the slit scan range. The fitting is satisfactory and each element of the **X** matrix is well determined. Figure 13 shows the two-dimensional distribution of **X** over the CCD at the scan center. Representative matrices for the left and right CCDs consisting of the median values of each element are also shown. Figure 14 shows **X**-matrix elements (median value in the CCD) as a function of the scan position. The horizontal dotted lines show the tolerance of each element defined in Section 2. From both Figures 13 and 14, we may regard the **X** matrix as uniform over both the CCD and the scan range (*i.e.*, within the tolerance defined in Section 3). There is, however, a systematic trend in some elements with the scan position or position in the CCD, which may be corrected in data reduction.

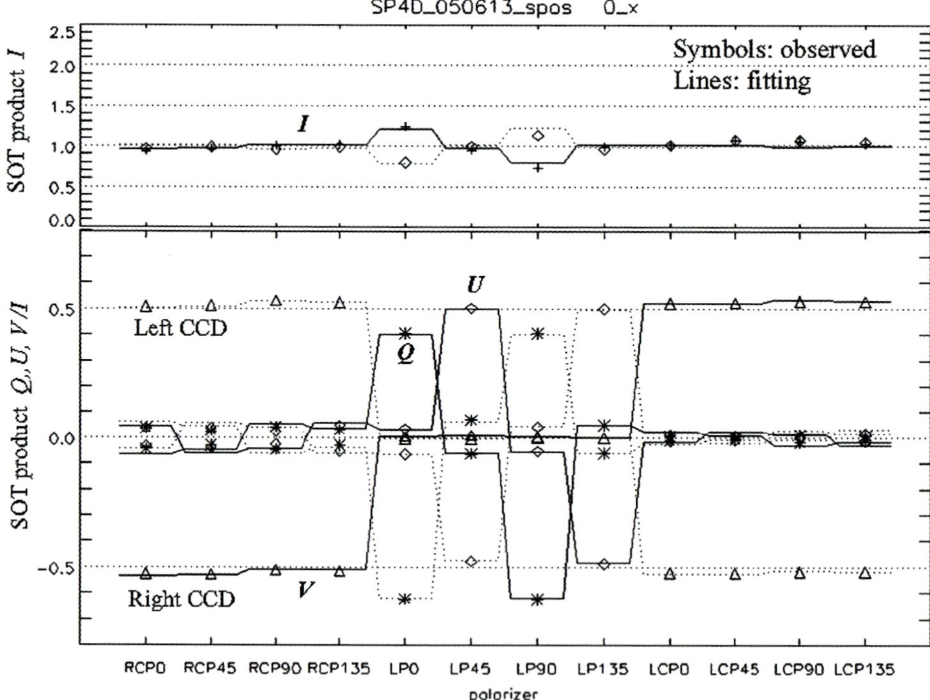

Figure 12 The observed SP products for 12 different configurations of the sheet polarizer (symbols) and results of the least-squares fitting (lines). The left and right CCDs are shown by solid and dotted lines, respectively.

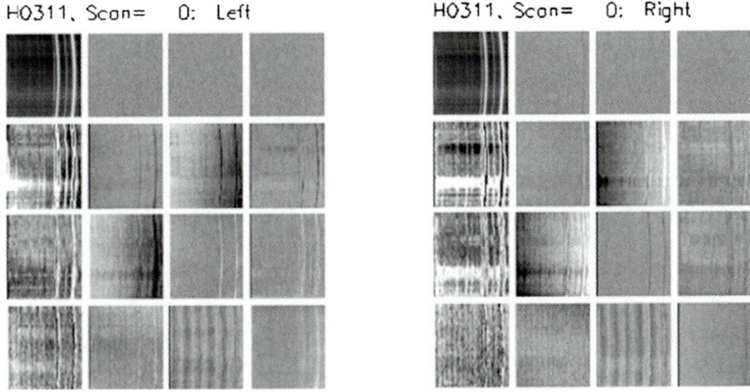

Median Mueller matrix

Left
```
 1. 0000  -0. 2232  -0. 0142  -0. 0063
 0. 0028  -0. 4819  -0. 0642   0. 0007
 0. 0022  -0. 0529   0. 4814  -0. 0030
-0. 0034  -0. 0026   0. 0043   0. 5249
```

Right
```
 1. 0000   0. 2077   0. 0199  -0. 0079
-0. 0039   0. 4886   0. 0551   0. 0005
-0. 0021   0. 0427  -0. 4918   0. 0034
 0. 0035   0. 0013  -0. 0044  -0. 5304
```

Figure 13 **X**-matrix spatial distribution over the CCD. Each element is scaled to median $\pm$ tolerance, and x_{00} ($=1$) is replaced by the I image.

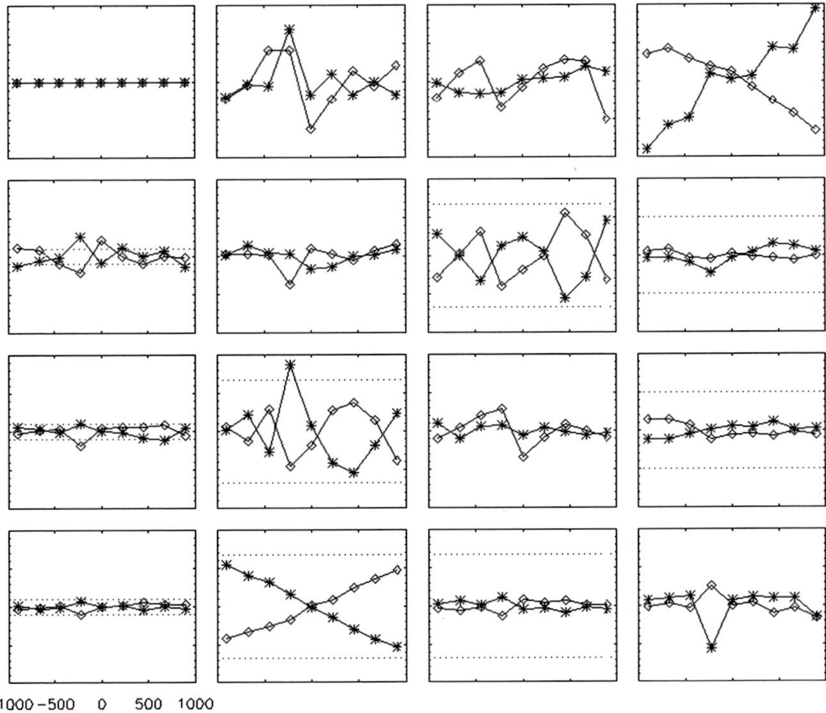

Figure 14 **X**-matrix elements (median value in CCD) as a function of scan position. Left and right CCD areas are shown by diamonds and asterisks, respectively. The horizontal dotted lines indicate the tolerance of each element as defined in Section 3.

5.5. NFI Polarimeter Response Matrix

Figure 15 shows examples of the observed products and results of the least-squares fitting for the NFI at 630.2 and 525.0 nm at the center of the CCD in the same manner as just demonstrated for the SP. The fitting is again successful, and the 15 elements of the response matrix are well determined for each CCD pixel. Figure 16 shows the spatial distribution of the **X** matrix over the CCD in NFI shuttered mode at 630.2 nm. The FOV covers the entire CCD, but the very left and right edges were obscured by an improper optical baffle during this test period in 2004. The plot in the right panel shows the horizontal cross sections of each element (with all rows overplotted). These plots again demonstrate that the **X** matrix may be regarded as uniform over the CCD. Figure 17 shows the spatial distribution of the **X** matrix over the CCD in NFI shutterless mode at 630.2 nm. The FOV covers a central 160×2048 pixel region. Note in the right-panel plot that the $(2, 3)$ and $(3, 2)$ elements show a discrete jump at the center of the FOV with an amount exceeding our tolerance. This behavior is due to the successive readout scheme of the CCD; the left half is read first and then the right half is read, and there is a time delay in effective exposure between left and right halves by about 2.83 ms. This jump corresponds to the mutual rotation of the $Q - U$ frame or rotation of the B azimuth between the two halves of the CCD by 2.83°. This effect will be corrected by the calibration. In each half of the CCD, the **X** matrix may be regarded as uniform. The experimental polarimeter response matrices were obtained for all wavelengths in which the NFI performs polarization measurements, namely, for 517.2, 525.0, 589.6, 630.2, and 656.3 nm.

5.6. Repeatability of the Measurement

To confirm the reliability of the measurements, we repeated the same measurement on different days during the test periods and checked the repeatability of the results. The following matrices show the difference of the SP **X** matrices (median for each left and right CCD) on two successive days (2005.6.13 and 6.14) as an example:

Left

0.0000	−0.0041	−0.0032	−0.0008
−0.0023	−0.0093	0.0021	0.0007
−0.0014	0.0040	−0.0071	−0.0008
−0.0012	0.0002	0.0001	0.0066

Right

0.0000	0.0368	0.0018	−0.0032
0.0094	0.0079	0.0048	−0.0069
0.0004	−0.0037	0.0099	−0.0000
0.0005	0.0006	−0.0014	−0.0053

The differences shown without underlining are within our tolerance, and we conclude that the accuracy of the measurement is good enough for most **X**-matrix elements except for the first column. Elements in the first column will be determined in orbit by using the continuum in the solar spectrum. This is also the case for the NFI.

6. Modeling of the SOT Polarization

We have successfully determined the experimental polarimeter response matrices for typical NFI observing sequences, but the NFI has a variety of exposure schemes with a variety of exposures, on-chip summing, and polarization sampling. Furthermore, new exposure schemes can be added after the launch whenever they are demanded. We do not have experimental **X**-matrix measurements for each case. To extend our knowledge of the **X** matrices of the

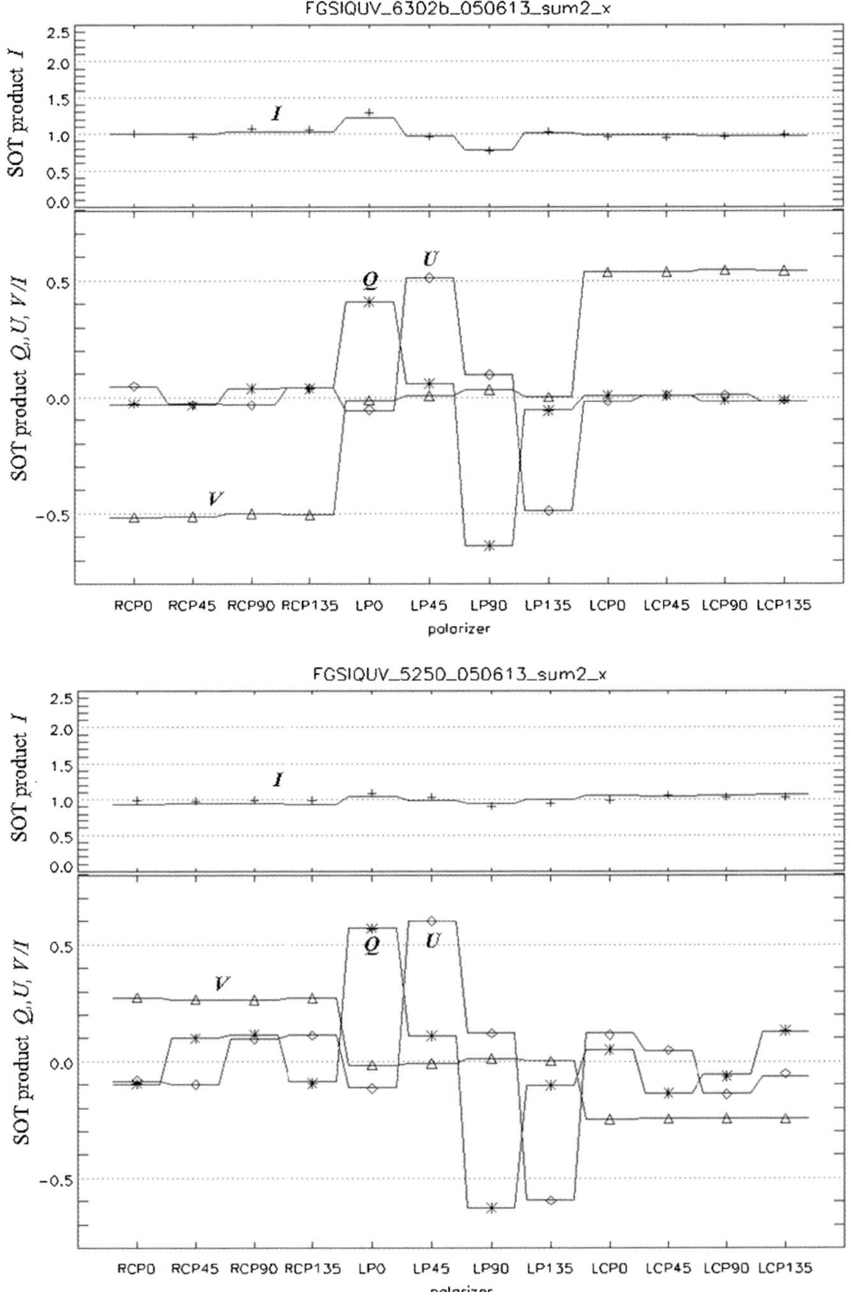

Figure 15 Same as Figure 12 but for the NFI at 630.2 nm (top) and 525.0 nm (bottom).

tested cases, a simple "SOT polarization model" is created, from which we may predict the **X** matrix for arbitrary observing sequences. There is another need for the SOT polarization model. Since, during the polarization calibration test, the entire FOV is illuminated by

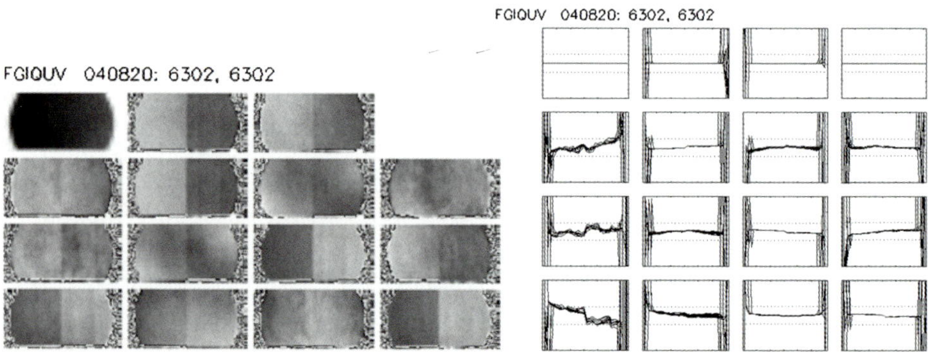

Figure 16 **X** matrix in NFI shuttered mode at 630.2 nm. Two-dimensional image over the CCD (left) and their plots against the *y*-coordinate (right). In the right panel, horizontal dotted lines show the tolerance of each element.

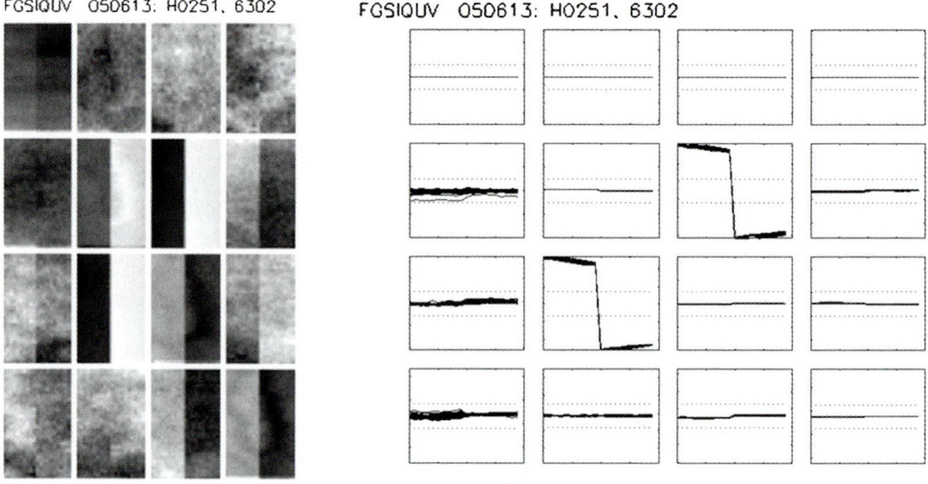

Figure 17 **X** matrix in NFI shutterless mode at 630.2 nm. Two-dimensional image over the CCD (left) and their plots against the *y*-coordinate (right). In the right panel, horizontal dotted lines show the tolerance of each element. The FOV covers the central 160×2048 pixels with 2 by 2 summing.

uniformly polarized light, some experimental **X**-matrix elements in shutterless mode have a variation across the FOV as the result of "polarization smearing" (see Section 6.2). Our SOT polarization model incorporates the correction of this artificial effect in experimental **X** matrices.

6.1. Basic Formulations

The assumptions of this model are as follows:

- The PMU is an ideal retarder and the polarization analyzer is an ideal linear polarizer.
- Exposure time and intervals between successive exposures are per flight software specifications.

- Any deviations of $\mathbf{X}$ from the theoretical matrix are attributed to the telescope matrix described in the following.
- The $\mathbf{X}$ matrix is uniform over each area (left and right) of the CCD.

The SOT data product, $\mathbf{S}_{\text{SOT}}$, and the polarimeter response matrix, $\mathbf{X}$, may be expressed as

$$\mathbf{S}_{\text{SOT}} = \mathbf{X}\mathbf{S}_{\text{in}} = \mathbf{DWTS}_{\text{in}},$$

or

$$\mathbf{X} = \mathbf{DWT},$$

where $\mathbf{D}$ is the demodulation matrix ($N \times 4$ elements, with N the number of exposures),

$$\mathbf{W} = (1, 1, 0, 0)\mathbf{P}(\delta_\lambda, \phi_k, \Delta t)$$

is the polarization measurement matrix ($4 \times N$ elements with $k = 0, \ldots, N-1$ standing for the sequence number of exposure), and $\mathbf{T}$ is the telescope Mueller matrix, with $(1, 1, 0, 0)$ as the first row of a Mueller matrix for an ideal analyzer and $\mathbf{P}(\delta, \phi_k, \Delta t)$ as the PMU matrix under continuous rotation, $k = 0, \ldots, N-1$, given by

$$\mathbf{P}(\delta_\lambda, \phi_k, \Delta t) = \int_{\phi_1}^{\phi_2} \mathbf{R}(-\phi)\mathbf{M}_{\text{ret}}(\delta_\lambda)\mathbf{R}(\phi) \, d\phi$$

$$= \begin{pmatrix} \Delta\phi & 0 & 0 & 0 \\ 0 & c_2 + s_2\cos\delta & (1-\cos\delta)d & -s_1\sin\delta \\ 0 & (1-\cos\delta)d & c_2\cos\delta + s_2 & c_1\sin\delta \\ 0 & s_1\sin\delta & -c_1\sin\delta & \Delta\phi\cos\delta \end{pmatrix}_k$$

where

$$c_1 \equiv \int_{\phi_1}^{\phi_2} \cos 2\phi \, d\phi = \left[\frac{1}{2}\sin 2\phi\right]_{\phi_1}^{\phi_2}, \qquad c_2 \equiv \int_{\phi_1}^{\phi_2} \cos^2 2\phi \, d\phi = \left[\frac{\phi}{2} + \frac{1}{8}\sin 4\phi\right]_{\phi_1}^{\phi_2},$$

$$s_1 \equiv \int_{\phi_1}^{\phi_2} \sin 2\phi \, d\phi = \left[-\frac{1}{2}\cos 2\phi\right]_{\phi_1}^{\phi_2}, \qquad s_2 \equiv \int_{\phi_1}^{\phi_2} \sin^2 2\phi \, d\phi = \left[\frac{\phi}{2} - \frac{1}{8}\sin 4\phi\right]_{\phi_1}^{\phi_2},$$

$$d \equiv \int_{\phi_1}^{\phi_2} \cos 2\phi \sin 2\phi \, d\phi = \left[-\frac{1}{4}\cos^2 2\phi\right]_{\phi_1}^{\phi_2}, \qquad \Delta\phi \equiv \phi_2 - \phi_1 = \omega \cdot \Delta t,$$

with

$$\phi_1 = \phi_k - \omega \cdot \Delta t/2,$$
$$\phi_2 = \phi_k + \omega \cdot \Delta t/2,$$
$$\omega = 2\pi/1.6 \text{ rad s}^{-1},$$

and where ϕ_k is the phase angles of the PMU at the center of each exposure, Δt is the exposure time, δ_λ is the retardation of the waveplate, and λ is the wavelength.

Using the response matrices obtained by the experiments, $\mathbf{X}_{\text{ex}}$, we performed a least-squares fitting for $\mathbf{X}$ with dt (exposure delay time) and δ_λ (retardation) as unknowns,

$$\mathbf{X}_{\text{ex}} \approx \mathbf{X}_{\text{fit}} = \mathbf{DW}(\phi_k, \Delta t, dt, \delta_\lambda)$$

with ϕ_k and Δt fixed to the specified values. The telescope matrix is determined by

$$\mathbf{T}(\lambda) = \mathbf{X}_{\text{fit}}^{-1}\mathbf{X}_{\text{ex}}$$

for each experimental data set. Then $\mathbf{T}(\lambda)$ and δ_λ are averaged over data sets for the same wavelength. The time delay dt is averaged for each left and right half of the CCD of the shuttered and shutterless modes. Thus the SOT polarization model provides the $\mathbf{X}$ matrix for arbitrary sequence specified by ϕ_k and Δt by

$$\mathbf{X}_{\text{model}} = \mathbf{DW}\big(\phi_k, \Delta t, \overline{\delta_\lambda}, \overline{\mathrm{d}t}\,\big)\mathbf{T}_\lambda.$$

6.2. Correction for Polarization Smearing

Each pixel in the SP and in the FG shutterless mode experiences smearing periods during the frame transfer, The mean PMU phase angle in an exposure (or effective exposure timing) differs with the position of the pixel on the CCD. Since the incident light is uniformly polarized in the FOV in the polarization calibration test, the variation of effective timing along the CCD appears as the slope of $(2, 3)$ and $(3, 2)$ elements of the derived response matrix along the CCD x-coordinate (see Figures 13 and 17). This is not the case for a real observation of the highly structured Sun. To eliminate this artificial effect in the experimental $\mathbf{X}$, we need to apply an additional correction to the SOT polarization model.

Figure 18 shows a chart describing the "exposure" sequence in NFI shutterless mode. In the chart, t_p is the center of the illuminated period $(\Delta t_0 + \Delta t_1 + \Delta t_2)$ whose timing is determined from the fitting of the experimental $\mathbf{X}$ matrix (dt). As inferred from the chart, this timing depends on the mask size and pixel position on the CCD. The parameter t_c is the center of the exposure cycles, which does not depend on either the mask size or pixel position.

The SOT data product is a summation of contributions from three periods – Δt_0, Δt_1, and Δt_2 – where Δt_0 is the period for charge being transferred from the CCD center to the pixel position x_p, Δt_1 is the period of the exposure at the pixel position, and Δt_2 is the period

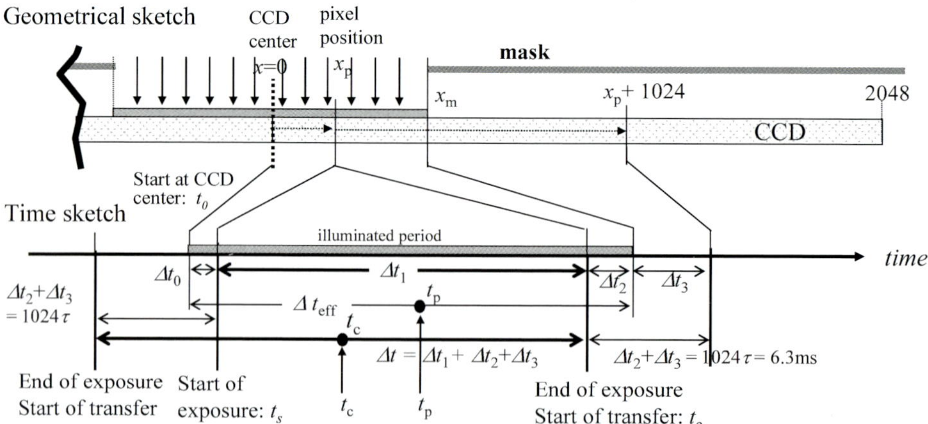

Figure 18 Geometrical and temporal sketch of exposure timing for the shutterless mode. Δt is the "exposure" time (typically $= 100$ ms); Δt_1 is the exposure at the pixel position; and Δt_0, Δt_2 is the smearing period.

for the charge being transferred from the pixel position x_p to the mask edge. Thus the SOT product is

$$\mathbf{S}_{\text{SOT}} = \mathbf{D}\big[\mathbf{W}(\Delta t_0)\mathbf{T}\,\mathbf{S}_{\text{in0}} + \mathbf{W}(\Delta t_1)\mathbf{T}\,\mathbf{S}_{\text{in1}} + \mathbf{W}(\Delta t_2)\mathbf{T}\,\mathbf{S}_{\text{in2}}\big].$$

In the polarization calibration test where the polarization is uniform over the SOT FOV,

$$\mathbf{S}_{\text{in0}} = \mathbf{S}_{\text{in2}} = \mathbf{S}_{\text{in1}},$$

and thus

$$\mathbf{S}_{\text{SOT}} = \mathbf{D}\big[\mathbf{W}(\Delta t_0) + \mathbf{W}(\Delta t_1) + \mathbf{W}(\Delta t_2)\big]\mathbf{T}\mathbf{S}_{\text{in1}}$$

and

$$\mathbf{X}_{\text{ex}} = \mathbf{X}(\Delta t_0) + \mathbf{X}(\Delta t_1) + \mathbf{X}(\Delta t_2).$$

In real solar observations, in contrast, since the smear regions (exposed areas during Δt_0 and Δt_2) are likely to have mixed polarity, Q, U, $V = 0$ on average and

$$\mathbf{S}_{\text{in0}} = \mathbf{S}_{\text{in2}} = (I, 0, 0, 0)^{\text{T}}.$$

Then

$$\mathbf{S}_{\text{SOT}} = (\Delta I, 0, 0, 0)^{\text{T}} + \mathbf{D}\mathbf{W}(\Delta t_1)\mathbf{T}\mathbf{S}_{\text{in1}} = (\Delta I, 0, 0, 0)^{\text{T}} + \mathbf{X}(\Delta t_1)\mathbf{S}_{\text{in1}},$$

where $\Delta I = I(\Delta t_0 + \Delta t_2)/(\Delta t_0 + \Delta t_1 + \Delta t_2)$ is the bias intensity resulting from smearing (with $\mathbf{T} \approx 1$ assumed). Since our aim is to obtain $\mathbf{S}_{\text{in1}}$ for a real observation,

$$\mathbf{S}_{\text{in1}} = \mathbf{X}(\Delta t_1)^{-1}\big[\mathbf{S}_{\text{SOT}} - (\Delta I, 0, 0, 0)^{\text{T}}\big], \quad \mathbf{X}(\Delta t_1) = \mathbf{D}\mathbf{W}(\Delta t_1)\mathbf{T}.$$

$\mathbf{X}(\Delta t_1)$ is the polarization calibration goal in NFI shutterless mode. As inferred from Figure 18, $\mathbf{X}_{\text{ex}}$ depends on both mask size and pixel position on the CCD, but $\mathbf{X}(\Delta t_1)$ is independent of both mask size and pixel position.

The target matrix $\mathbf{X}(\Delta t_1)$ may then be calculated by the following equation:

$$\mathbf{X}(\Delta t_1) = \int_{t_s}^{t_e} \mathbf{X}(t)\,dt = \int_{t_c - \Delta t/2 + 1024\tau}^{t_c + \Delta t/2} \mathbf{X}(t)\,dt,$$

where t_c (the center of the exposure cycle, which is independent of the pixel position and the mask) is converted from t_p by

$$t_c = t_p - \tau(512 + x_m/2 - x_p),$$

with $\tau = 0.00615$ ms/pixel being the rate of frame transfer. Using this formula, one can correct the exposure delay time dt in the polarization model.

6.3. Results from the SOT Polarization Model

Table 5 summarizes the parameters obtained for the SOT polarization model for each wavelength. The integer part of the retardation of the PMU is that specified in the design of the waveplate. Notice that the modulation amplitudes for both linear and circular polarization

Table 5 Parameters of the SOT polarization model.

| Wavelength (nm) | Retardation (waves) | | Modulation amplitude (diagonal element of X in shutterless mode) | | Time delay of t_c (ms) | | | |
| | | | | | Shutterless | | Shuttered | |
	Design	Measured	QU	V	Left	Right	Left	Right
517.3	6.650	6.6822	0.45	0.58	−0.24	6.16	–	–
525.0	6.558	6.5720	0.61	0.27	0.80	7.09	−5.52	−5.55
589.6	5.816	5.7624	0.30	0.63	0.28	6.63	−5.47	−6.05
630.2	5.350	5.3442	0.50	0.53	−1.47	4.93	−7.99	−7.52
656.3	5.050	5.1095	0.07	0.40	−4.23	3.02	−9.87	−9.35

Note. There is no measurement of the time delay in the shuttered mode at 517.3 nm. In the reduction of flight data, the same delay at 525.0 nm will be assumed.

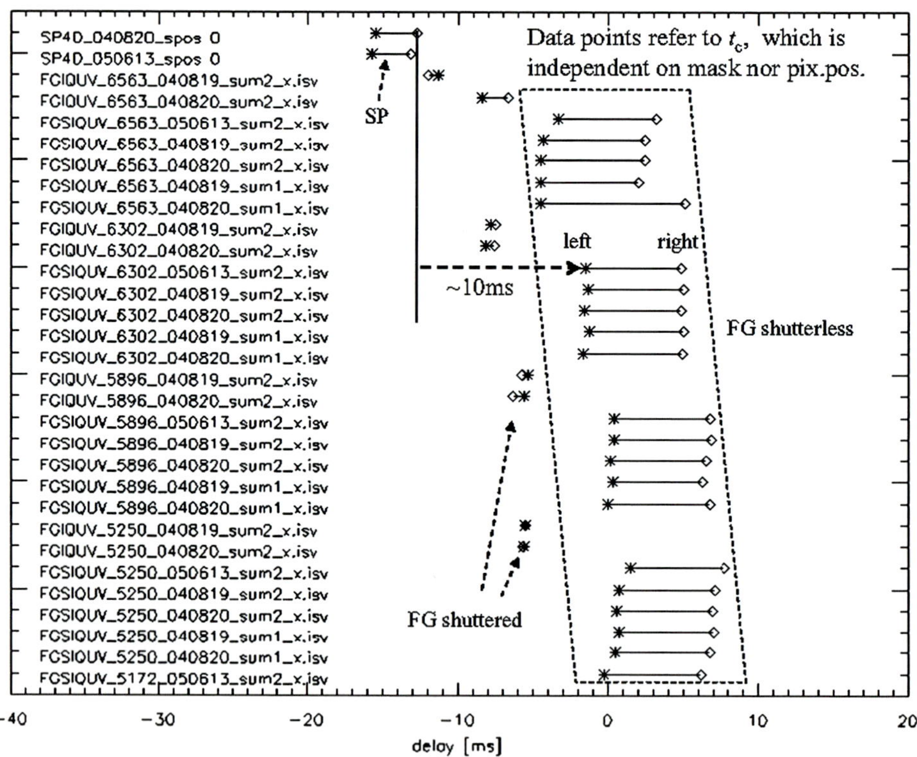

Figure 19 Time delay of exposure in all data sets. Asterisks and diamonds refer to the left half and right half, respectively, of the CCD.

are equivalent at 517.3 and 630.2 nm (as intended by the design) and high modulation efficiency for circular polarization at 589.6 nm, high modulation efficiency for linear polarization at 525.0 nm, and less modulation efficiency for linear polarization in Hα are realized. In Figure 19, a plot of time delay (effective exposure timing) for all data sets used in the analysis is shown. There is an obvious dependence of the delay on wavelength, the cause of

Table 6 Telescope matrices and the standard deviation of the fitting residual with the SOT polarization model.

Average T matrix				STD deviation of fitting residual			
6563	0.9893	−0.0420	−0.0491	0.0018	0.0000	0.0117	0.0296	0.0113
	−0.0121	−0.9541	0.0190	0.0072	0.0006	0.0025	0.0005	0.0009
	−0.0052	0.0088	0.9764	0.0205	0.0010	0.0011	0.0015	0.0013
	−0.0049	−0.0285	−0.0135	1.0070	0.0008	0.0010	0.0019	0.0067
6303	0.9976	0.0101	0.0276	0.0031	0.0000	0.0069	0.0087	0.0062
	0.0108	0.9990	0.0145	−0.0025	*0.0028*	0.0080	0.0022	0.0036
	0.0030	0.0131	0.9983	−0.0157	*0.0012*	0.0021	0.0079	0.0017
	−0.0050	0.0437	0.0099	0.9763	*0.0015*	0.0020	0.0010	0.0086
5896	0.9951	0.0008	0.0730	−0.0006	0.0000	0.0075	0.0107	0.0028
	0.0091	0.9970	0.0144	−0.0010	*0.0018*	0.0046	0.0018	0.0013
	0.0013	0.0147	1.0021	−0.0143	*0.0018*	0.0016	0.0049	0.0019
	−0.0099	−0.0178	0.0111	0.9927	0.0010	0.0031	0.0009	0.0103
5250	0.9994	0.0061	0.0141	−0.0082	0.0000	0.0040	0.0148	0.0199
	0.0113	0.9996	0.0131	0.0011	*0.0032*	0.0083	0.0046	0.0027
	0.0030	0.0136	1.0031	−0.0169	*0.0043*	0.0033	0.0137	0.0042
	−0.0169	−0.0459	0.0025	0.9931	*0.0011*	0.0015	0.0020	0.0074
5172	0.9998	0.0007	−0.0296	−0.0458	–	–	–	–
	−0.0007	1.0003	0.0077	−0.0077	–	–	–	–
	−0.0018	0.0093	0.9863	0.0149	–	–	–	–
	−0.0139	0.0543	−0.0246	0.9901	–	–	–	–

Note. Elements in italics exceed the tolerance. Standard deviations are not given for 5172 Å since we have only a single measurement for this wavelength.

which is not well understood but is likely due to fabrication error of the waveplate. Residuals of the fitting (*i.e.*, the difference of $\mathbf{X}_{ex}$ and $\mathbf{X}_{fit} = \mathbf{DW}$) averaged in each wavelength for the telescope matrices are shown in Table 6. The standard deviations of the fitting residual are also shown in the right side of Table 6. Since all elements of the standard deviation other than the first column are smaller than the tolerances of the response matrix, we can consider that the SOT polarization model developed here well represents the real polarization response matrix of the NFI except for the first column. The elements of the first column will be determined in orbit, again by using continuum values.

For reference, the polarimeter response matrices provided by the SOT polarization model are presented in Table 7 for the NFI shutterless *IQUV* mode, where the telescope matrix is assumed to be unity.

7. Examples of Extended Observing Schemes for the NFI

7.1. Magnetogram (*IV* Mode)

So far we have been focusing on observations of full Stokes parameters in which a 4×4 polarimeter response matrix is applicable to retrieve the incident Stokes vector. The NFI can

Table 7 The polarimeter response matrices provided by the SOT polarization model for the NFI shutterless *IQUV* mode (Obs_ID = 33, exposure = 100 ms). The telescope matrix is assumed to be unity.

	Left-CCD				Right-CCD			
6563	1.0000	0.8863	0.0000	0.0000	1.0000	0.8863	0.0000	0.0000
	0.0000	0.0723	0.0048	0.0000	0.0000	0.0723	−0.0034	0.0000
	0.0000	0.0048	−0.0723	0.0000	0.0000	−0.0034	−0.0723	0.0000
	0.0000	0.0000	0.0000	−0.4040	0.0000	0.0000	0.0000	−0.4041
6303	1.0000	0.2210	0.0000	0.0000	1.0000	0.2210	0.0000	0.0000
	0.0000	0.4958	0.0114	0.0000	0.0000	0.4944	−0.0384	0.0000
	0.0000	0.0114	−0.4958	0.0000	0.0000	−0.0384	−0.4944	0.0000
	0.0000	0.0000	0.0000	−0.5279	0.0000	0.0000	0.0000	−0.5279
5896	1.0000	0.5389	0.0000	0.0000	1.0000	0.5389	0.0000	0.0000
	0.0000	0.2935	−0.0013	0.0000	0.0000	0.2919	−0.0305	0.0000
	0.0000	−0.0013	−0.2935	0.0000	0.0000	−0.0305	−0.2919	0.0000
	0.0000	0.0000	0.0000	0.6374	0.0000	0.0000	0.0000	0.6338
5250	1.0000	0.0503	0.0000	0.0000	1.0000	0.0503	0.0000	0.0000
	0.0000	0.6046	−0.0076	0.0000	0.0000	0.6009	−0.0672	0.0000
	0.0000	−0.0076	−0.6046	0.0000	0.0000	−0.0671	−0.6009	0.0000
	0.0000	0.0000	0.0000	0.2783	0.0000	0.0000	0.0000	0.2778
5172	1.0000	0.2934	0.0000	0.0000	1.0000	0.2934	0.0000	0.0000
	0.0000	0.4498	0.0017	0.0000	0.0000	0.4477	−0.0434	0.0000
	0.0000	0.0017	−0.4498	0.0000	0.0000	−0.0434	−0.4477	0.0000
	0.0000	0.0000	0.0000	0.5797	0.0000	0.0000	0.0000	0.5790

also take IV information only with two exposures centered at the PMU phases at $\pm 45°$ (Figure 20) and the exposure time is selectable in shuttered mode. In practice such a mode, called a "magnetogram," is useful for making context longitudinal magnetograms at high cadence.

Intensities obtained by the two exposures are given by

$$I_+ = I + c_Q Q + c_V V,$$

$$I_- = I + c_Q Q - c_V V$$

and the polarimeter response matrix in this case consists of 4×2 elements,

$$\begin{pmatrix} I' \\ V' \end{pmatrix} = \begin{pmatrix} x_{00} & x_{10} & x_{20} & x_{30} \\ x_{03} & x_{13} & x_{23} & x_{33} \end{pmatrix} \begin{pmatrix} I \\ Q \\ U \\ V \end{pmatrix} \approx \begin{pmatrix} 1 & c_Q & 0 & 0 \\ 0 & 0 & 0 & c_V \end{pmatrix} \begin{pmatrix} I \\ Q \\ U \\ V \end{pmatrix}.$$

Here c_Q and c_V represent $Q \rightarrow I$ crosstalk and the efficiency of the V measurement, respectively, and are functions of exposure time. Figure 21 shows the plots of c_Q and c_V against the exposure time predicted from the SOT polarization model for five NFI wavelengths. The verification test of such curves was performed by using the FPP and backup

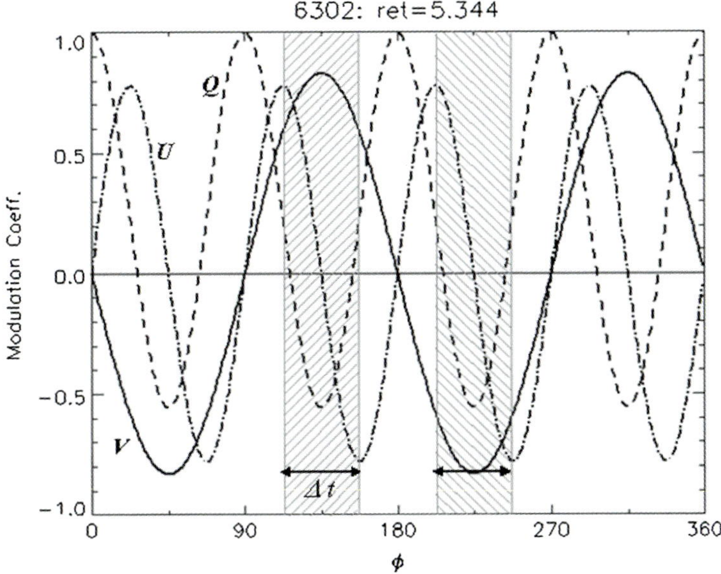

Figure 20   Exposure timing with respect to the PMU modulation phase for the NFI IV mode.

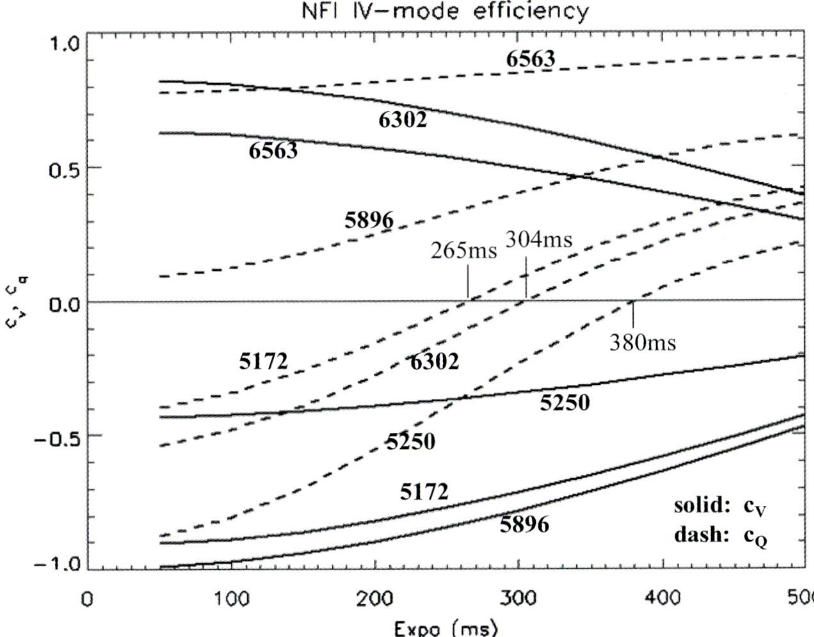

Figure 21   Plots of c_Q and c_V against the exposure time predicted from the SOT polarization model. Solid curves are c_V (efficiency of V measurement) and dashed curves are c_Q ($Q \rightarrow I$ crosstalk).

(flight spare) unit of the PMU. Figure 21 suggests that there is a preferable exposure time at which the $Q \rightarrow I$ crosstalk becomes negligible for each wavelength.

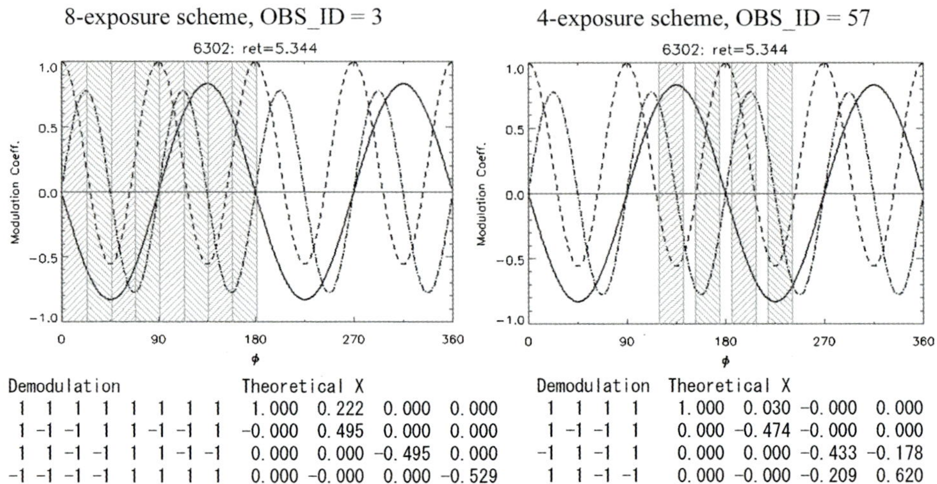

Figure 22 Eight-exposure scheme and four-exposure scheme of the NFI to get $IQUV$.

7.2. NFI Four-Exposure $IQUV$ Mode

In addition to the eight-exposure scheme to take $IQUV$, the FG shuttered mode has a four-exposure scheme whose **X** matrix was not measured with the real SOT. Figure 22 shows the exposure timing with respect to the PMU modulation curve for both schemes together with the theoretical polarimeter response matrices. The four-exposure scheme takes four exposures with equally spaced PMU phase angles centered at $180°$. In our case the space is set as $33.1°$ with which we can maximize the modulation efficiency for V and equalize the modulation efficiencies for Q and U. The verification test of this scheme was again performed by using the FPP and PMU flight spare unit at 630.2 and 517.2 nm, and we have confirmed that the SOT polarization model well predicts the polarimeter response matrix of the four-exposure scheme.

8. Summary

Polarimeter response matrices of the SOT were determined experimentally by using the entire SOT for the SP and typical NFI observing modes as functions of position in the FOV. Matrix elements are determined to sufficient accuracy so that crosstalk among different Stokes parameters is suppressed below the typical statistical noise of 10^{-3}. The accuracy of measurements inferred from repeatability meets the required accuracy except for the first column of the matrices. The polarimeter response matrices can be regarded as uniform over the field of view except for the NFI shutterless mode, in which the left and right halves of the CCD have a non-negligible difference owing to the relative exposure delay between the two halves. Even though the variation of **X** for the SP is also smaller than the tolerance, we detected a smooth variation of the **X** matrix of the SP across its field of view, which is taken into account for calibration of flight data. The SOT polarization model reproduces experimental **X** matrices of the NFI with the required accuracy and can be used to predict the **X** matrices of other observing sequences for which the experimental **X** matrices were not obtained. The SOT polarization characteristics are expected to be fairly stable in orbit, whereas the linear retardation of the CLU may have a small offset created during the launch load environment. This possible offset may be checked in real solar data and the $\mathbf{X}$ matrix will be updated if necessary.

Following the successful launch of *Hinode* on 22 September 2006, the SOT achieved its first light soon after the deployment of the top door of the telescope on 25 October. After the initial instrument checkout, the SP and the FG began making regular observations of the Sun and are producing excellent Stokes data. We have noticed that the three elements of the first column of $\mathbf{X}$ matrices are very close to zero from the SP and the FG/NFI continuum data. However, we have not yet found any sign implying an offset of the CLU linear retardation ($V \rightarrow QU$ crosstalk). We need to await further detailed analysis of sunspot data to finalize this issue.

Polarization calibrations of the SP data are being performed successfully on a regular basis by using the polarimeter response matrices obtained by HAO. An IDL procedure (fg_pcalx.pro) that provides polarization response matrices for arbitrary NFI products is ready and catalogued in the Solar Software (SSW) package.

Acknowledgements   The authors are grateful to the late professor T. Kosugi of JAXA/ISAS and to Drs. L. Hill, R. Jayroe, and J. Owens of NASA for continuous support throughout the development of the SOT. The authors also thank Messrs. T. Matsushita and H. Saito of Mitsubishi Electric Corporation and Mr. N. Takeyama of Genesia Corporation for extensive thermomechanical analysis of the CLU and also Messrs. S. Abe and M. Suzuki of Canon Corporation for indispensable support for the polarization testing of the CLU. One of the authors (KI) also would like to thank Dr. E. West of Marshall Space Flight Center, Dr. R. Chipman of Alabama University, and Dr. Y. Otani of Tokyo University of Agriculture and Technology for valuable discussions on the polarization measurements of optical devices in the early phase of the project. *Hinode* is a Japanese mission developed and launched by ISAS/JAXA, with NAOJ as domestic partner and NASA and STFC (UK) as international partners. It is operated by these agencies in cooperation with ESA and NSC (Norway).

References

Elmore, D.F.: 1990, *A polarization calibration technique for the advanced stokes polarimeter. NCAR Technical Note NCAR/TN-355+STR*, NCAR, Boulder, Colorado.

Guimond, S., Elmore, D.: 2004, *OE Mag.* **4**(5), 26.

Ichimoto, K., Tsuneta, S., Suematsu, Y., Shimizu, T., Otsubo, M., Kato, Y., et al.: 2004, In: Mather, J.C. (ed.) *Optical, Infrared, and Millimeter Space Telescopes, Proc. SPIE* **5487**, 1142.

Ichimoto, K., the Solar-B Team: 2005, *J. Korean Astron. Soc.* **38**, 307.

Ichimoto, K., Shinoda, K., Yamamoto, T., Kiyohara, J.: 2006, *Publ. Natl. Astron. Obs. Japan* **8**, 11.

Kosugi, T., Matsuzaki, K., Sakao, T., Shimizu, T., Sone, Y., Tachikawa, S., et al.: 2007, *Solar Phys.* **243**, 3.

Shimizu, T.: 2004, In: Sakurai, T., Sekii, T. (eds.) *The Solar-B Mission and the Forefront of Solar Physics* **CS-325**, Astron. Soc. Pac., San Francisco, 3.

Shimizu, T., Nagata, S., Edwards, C., Tarbell, T., Kashiwagi, Y., Kodeki, K., et al.: 2004, In: Mather, J.C. (ed.) *Optical, Infrared, and Millimeter Space Telescopes, Proc. SPIE* **5487**, 1199.

Shimizu, T., Nagata, S., Tsuneta, S., Tarbell, T., Edwards, C., Shine, R., et al.: 2008, *Solar Phys.* in press.

Skumanich, A., Lites, B.W.: 1997, *Astrophys. J. Suppl.* **110**, 357.

Suematsu, Y., Tsuneta, S., Ichimoto, K., Shimizu, T., Otsubo, M., Katsukawa, Y., et al.: 2008, *Solar Phys.* in press.

Tarbell, T.D., et al.: 2008, *Solar Phys.* to be submitted.

Tsuneta, S., Ichimoto, K., Katsukawa, Y., Shimizu, T., Otsubo, M., Nagata, S., et al.: 2008, *Solar Phys.* in press.